Springer Praxis Books

Popular Astronomy

The *Springer-Praxis Popular Astronomy* programme covers the latest observations, techniques and discoveries in astronomy, astrophysics and cosmology, selecting key topics designed to fire the imagination of keen astronomers everywhere. Many of the topics selected for titles in this programme are refreshingly new, and away from the accepted mainstream of titles published in this area.

Topics include:

- the search for water and life both within our Solar System and beyond
- major discoveries at the edge of the Solar System,
- extrasolar planets
- the history of our Galaxy
- galaxy collisions and mergers
- the origins and evolution of the Universe

The books are well illustrated with figures, photographs and maps, with extensive use of colour for scientific interpretation and understanding. They feature recommended further reading and glossaries and appendices where appropriate.

The books are written in a style that astronomy enthusiasts, readers of popular magazines such as *Sky and Telescope* and *Astronomy*, and readers of *Scientific American* will find accessible.

More information about this series at http://www.springer.com/series/7144

Gabriella Bernardi · Alberto Vecchiato

Understanding Gaia

A Mission to Map the Galaxy

 Springer

Gabriella Bernardi
Torino, Italy

Alberto Vecchiato
National Institute of Astrophysics (INAF)
Pino Torinese, Italy

ISSN 2626-8760   ISSN 2626-8779   (electronic)
Springer Praxis Books
Popular Astronomy
ISBN 978-3-030-11448-0   ISBN 978-3-030-11449-7   (eBook)
https://doi.org/10.1007/978-3-030-11449-7

Library of Congress Control Number: 2018967427

This Praxis imprint is published by the registered company Springer Nature Switzerland AG
The registered company address is: Gewerbestrasse 11, 6330 Cham, Switzerland

All nature is but art, unknown to thee;
All chance, direction, which thou canst not see;
All discord, harmony not understood;
All partial evil, universal good.
And, spite of pride, in erring reason's spite,
One truth is clear, "Whatever is, is right."
Alexander Pope, An Essay on Man

To our families

Preface

This book started out as the evolution of a previous little book that one of us (GB) wrote on the occasion of the launch of Gaia, a satellite built by the European Space Agency. Its purpose was to exploit the occasion of this space mission as an opportunity to write a short and introductory text on astronomy and space science for the layman, but, at the same time, keeping one eye toward young people.

The popularization of science is anything but a simple profession, and it often requires expertise in different fields, so that a short book also represented a nice chance for the author to dust off the competence in regard to space science that she had acquired years before, while working on the Rosetta satellite. In that case, such expertise was very useful, but the goal of the present book required a longer and more detailed exposition, for which collaboration with a scientist who has worked on the Gaia mission for several years constituted a valuable addition.

The Gaia satellite, launched at the end of 2013, is responsible for a major advance in astronomy and astrophysics, leveraging the simple concept of providing a complete and high-precision map of the positions and velocities of the stars in our galaxy. No matter how simple, this idea is boosted by the sheer power of the vastness of such a map, and by one of its fundamental ingredients: the stellar distances.

Thanks to these ingredients, Gaia will revolutionize our knowledge on the origin and evolution of the Milky Way, on the effects of mysterious Dark Matter, and on the birth and evolution of stars and extrasolar planets. In other words, its results will foster a total breakthrough, whose consequences will span all of the realms of astronomy and astrophysics.

This space mission, therefore, represents a wonderful opportunity to talk about several aspects of this science that are often treated separately. Among them, a special place in this book is reserved for astrometry. Despite its fundamental importance, this branch of astronomy, which deals with the measurement of stellar positions, is poorly covered in popular science publications, which tend to focus on other, more popular fields. A satellite like this, instead, represents the perfect chance to explain what astrometry is and how it works.

Gaia produces a steady flow of data that, during the course of its operational life, will yield a massive database of raw measurements. One of the main difficulties of the mission is the processing of said data, namely, the transformation of this impressive amount of unorganized data into an invaluable vault of scientific information. Thus, a specific portion of the book is devoted to explaining this complex process and its scientific consequences.

Building and operating a satellite is always a difficult task that requires mastering a lot of knowledge, from astronautics to information technology. Here, we also used Gaia to take a fairly detailed look at these more specialized aspects of a space mission.

Finally, although it will require several more years to obtain the satellite's results in their final form, a series of intermediate releases have already been published. The last part of the book, then, is devoted to showing and explaining a selection of them.

In summary, this work will take the reader on an all-inclusive journey of discovery between the skies and the science and technology of a scientific satellite, showing, at the same time, how the data are collected, interpreted, and used for scientific purposes. To this aim, we used information available from publicly accessible sources only, but the resulting work benefits from the combined forces of a professional science writer and a scientist who was able to exploit his insider's point of view. We also wish to thank the following people who gave us feedback on various parts of the book or helped us find images: Beatrice Bucciarelli, Roberto Morbidelli, Paolo Tanga, and Claudia Travaglio.

Torino, Italy   Gabriella Bernardi
October 2018   Alberto Vecchiato

Contents

Introduction

The year 1989 represents a historical moment for Astrometry. This is the year when Hipparcos, the first astrometric satellite, was launched, symbolizing the beginning of the modern era of astrometry.

Until then, in fact, this incredibly ancient and glorious branch of science had been undergoing a decline that lasted for more than a century, since the scientific and technological advances of the nineteenth century had given birth to and nurtured the development of Astrophysics.

Astrometry, intended as the way to measure the positions and motions of the stars in the sky, has accompanied humankind since its very beginning as a species. Some of the constellations, like the Big Dipper, Taurus, and others, are so familiar to us that it might seem as if they have existed forever. And this could be one of the cases in which our instinct is right, as some studies seem to have ascertained that they represent the heritage of an oral tradition that can be dated back to 17,000 years. In the Paleolithic era, well before the most ancient civilizations and even before human beings learned the art of writing, people felt a need to identify the stars in the sky, giving them names and making beautiful paintings of their stellar-inspired mythology in their caves.

In a certain sense, this cannot be surprising. Knowledge of the configuration of the stars and their motion told when the seeds could be planted, when the harvest could be gathered, or where the next hunting area was located. In other words, it represented the difference between life and death.

When the first civilizations appeared in history, the ability to predict astronomical events like eclipses added to this tradition, and the people who possessed these skills could interpret the will of the gods, or even command them!

With the advent and development of Astronomy as an exact science, Astrometry was the observational counterpart of Celestial Mechanics. These two scientific sisters constituted the entire realm of Astronomy. The systematic observations needed to obtain certain significant result required years of long, hard work, and a bit of luck, but such was the importance of discovering a new comet, or producing a precise stellar chart, that the reward could even change the life of the lucky, hard-working astronomer.

Then, Astrophysics completely subverted this scenario, and it happened so fast that, in a matter of just a few decades, Astrometry was left with no glory but just with the long and hard working. Suddenly, with the help of new astrophysical techniques, for the first time, scientists could understand what the stars were made of and how they evolved. They measured stellar temperatures and understood the secret of the incredible amount of energy within them. People were informed that stars assemble together into galaxies, and that galaxies group themselves into clusters and superclusters, and that the Universe was born about 13 billion years ago from a super-small, dense, and hot point, and is currently expanding at an incredible speed. Like a kind of lucky Pandora's Box, the coming of Astrophysics made a seemingly easy and unstoppable flux of new and incredible discoveries possible, along with the flourishing of new theories. Something that quickly dwarfed the labored achievements of a thousand of years-long tradition of Astronomy.

Positional astronomy continued to be practiced, but it compared to Astrophysics like a cornfield compared to the Klondike during the roughly contemporary Gold Rush, so the leverage had definitely and permanently moved elsewhere… or had it?

Actually, Klondike prospectors could not survive without corn, and, similarly, Astrophysics needed astrometric data to feed and sustain its wild development. Soon, it became clear that Astrophysicists without good astrometric data were like a swarm of explorers without a good map. Surely, each of them could keep up exploration at an impressive speed, but the farther one moves, the larger the unexplored territory, and the longer it takes to find the right direction if one does not have the right tools.

This comparison is more literal than it might seem, since it was a matter of maps, indeed. At the very end, one of the most fundamental pieces of information needed to answer the uncountable questions raised by astronomical research is the distance of the objects under investigation. There are several techniques for estimating astronomical distances, each with its own range of validity. Moreover, they depend on each other with a general schema. Those applicable to smaller distances fail when the object is too far away, but, on the other hand, you need to finely tune the large-distance techniques to make them useful; otherwise, their accuracy is too poor. This fine-tuning is provided by each previous method in the series; therefore, you need to build this so-called cosmic distance ladder step by step to reach the farthest objects in the universe. And guess what the first of these rungs is: the parallax determination, which is the classical astrometric technique for measuring distances!

In principle, this is a purely geometric method, and it is self-consistent, but the parallax, which relies on the measurement of angle differences, is also extremely difficult to estimate. Moreover, there are two types of parallax, relative and absolute, and, although both are useful, the latter is the "philosopher's stone" of the cosmic distance ladder. You can bet that it is thus also the most difficult to obtain.

Determining the absolute parallax is an endeavor that deals with the realization of maps of the entire sky, something that astrometrists call "global catalogues," which is the most difficult of their endeavors. In fact, as late as the second half of the last century, the best global catalogues counted a few thousand objects, and only a handful of the nearest stars had a reliable relative parallax.

A large share of the difficulties came from the fact that, for various reasons, the necessity of observing from the ground limited the accuracy and speed of these measurements, which is why Hipparcos came into play. As a space-based measuring instrument, it made possible the realization of a catalogue 100 times larger and more accurate than its most recent predecessors, and it did it in about thirty years (a period that goes from the initial idea to the release of the final catalogue, and includes just 2.5 years of measurements).

Nevertheless, Hipparcos' horizon was limited to a small region around the Sun, some 100 times smaller than our Galaxy, in linear distance. During the last years of this mission, like a symbolic handover, a more ambitious program was conceived, that of another satellite that would have been able to measure a far greater number of stars and with much better accuracy. Something that could allow, for the first time in history, the realization of a three-dimensional map (actually, five-dimensional, but this will be revealed later) of practically the entire Milky Way. Its name was Gaia, and its promise is doing nothing less than revolutionizing our astronomical knowledge and keeping astronomers busy with its data for the next several decades.

Like all bird's-eye views, this can give us more or less of a general picture, but leaves most questions unanswered. Why, precisely, have hundreds of astronomers planned to work so hard for decades (Gaia will last for about thirty years, like Hipparcos) to realize a map of the sky? Who made the first star chart? How are astronomical satellites built and operated? How do they communicate with us, and how are their data processed to produce the final map? What will happen afterward? And, last, but not least … what the heck is the Milky Way?!

Gaia was launched in 2013 and, to date, the European Space Agency and the consortium of scientists that was established to process its data has released the second in the series of its catalogues, while the final one is not expected before 2023. The goal of this book is to try to answer these and other questions, and to show and explain the initial results of the mission. In a certain sense, the satellite itself, with its name, will guide us on this route, which could therefore rightly be called "The Spelling of Gaia."

Part I
G as in Gaia, Galaxy, Galaxies

We shall not cease from exploration
And the end of all our exploring
Will be to arrive where we started
And know the place for the first time.

T. S. Elliot, Four Quartets

1

Gaia in the Sky

The first thing that is necessary when studying the sky is direct observation of it. In the past, throughout the world, the ancients began with our primary tools: the eyes. Then came astrolabes, lenses, and bigger and bigger telescopes, protected by even bigger domes, until the recent introduction of satellites, able to capture light in different bands of the electromagnetic spectrum, completing the present picture. In parallel with the technological development of the instrumentation, humanity has developed strategies and models for understanding what it observes.

In this book, we want to tell a recent and still unfinished episode of this never-ending story; one starring a satellite named Gaia, combining a modern rendition of one of the most ancient sciences—astrometry—with extremely advanced technologies and sophisticated scientific models. A scientific endeavor that pursues the anything-but-humble promise of revolutionizing every single aspect of astronomy and astrophysics.

The very name of this satellite, Gaia, has an interesting story by itself. As often happens with big projects, it was born as the all-capitals acronym GAIA, which stood for Global Astrometric Interferometer for Astrophysics. The "I" referred to the kind of measuring instrument that was originally thought to be housed in the space vehicle, however, in the final version of the project, the interferometer disappeared, meaning the acronym no longer made sense. The name, however, remained very fitting.

Because of scientific concepts like the Gaia hypothesis, the common parlance of today simply identifies Gaia with the Earth, or, at most, with some more ecologically-biased "Mother Earth" (Fig. 1.1). Yet, in antiquity, this name used to have a deeper and more fundamental meaning. In ancient Greek mythology, Gaia is the name of a primordial goddess; one that came after the god Chaos—the infinite void or the shapeless matter and the first existing thing—and who then created the god of the heavens, Uranus, the gods of the sea and the mountains, and

© Springer Nature Switzerland AG 2019
G. Bernardi and A. Vecchiato, *Understanding Gaia*,
Springer Praxis Books, https://doi.org/10.1007/978-3-030-11449-7_1

Fig. 1.1 Floor mosaic from a Roman villa in Sassoferrato, Italy dated about 200 A.D. It represents Aion, the Hellenistic god of time, and the Roman Gaia, mother Earth

all of the other gods of Mount Olympus, including Kronos, the god of time. She is thus the equivalent of the creation and—pun intended—she brought order after chaos, making the world understandable in human terms.

If this is not a proper name for a mission whose goal is to understand the origin and evolution of our galaxy, and, in general, to "bring order after chaos" by revolutionizing our understanding of the heavens, nothing else can be!

Gaia was built in Europe by a consortium of industries and scientists led by ESA, the European Space Agency, and it was launched on December 19, 2013; yet, as often happens with satellites, it had been conceived many years before, in 1993, when another similar mission was coming to its conclusion. Gaia's predecessor was the High-Precision Parallax Collecting Satellite, or HiPParCoS, another appropriate choice for the first astrometric satellite that, with its name, celebrated the astrometrist who determined the first parallax in history, that of the

Moon (see Chap. 5). In 1992, the Hipparcos satellite, launched in its turn in 1989 by ESA, had just ended its operational phase, and, while the delicate task of processing the data was still running full steam ahead, scientists started to think of the successor to this mission. One that could exploit the newest and most advanced technologies so as to boost Hipparcos's goal of mapping the stars in "space and time" by orders of magnitude. Suddenly, Gaia was born. In its DNA were written instructions to make a genuine quantum leap from the already great achievements of its forerunner possible. If Hipparcos could map a few hundred-thousandths of the Milky Way, Gaia would have been able to cover it all over its entire extension, and with an accuracy more than 100 times better; if the stars in the previous map were on the order of 100,000, now they would be counted in the hundreds of millions!

But wait. Stars, the Milky Way, Galaxies… Actually, before venturing into this quest, it is wise to make sure that we have the knowledge very much needed to understand this enterprise. We must know what the Milky Way and other similar galaxies are, maybe together with their building blocks, the stars.

2

The Milky Way: Our Galaxy and Its Components

2.1 The Galaxy

On a clear night, and from a place free from light pollution, it is impossible not to notice a bright, imposing strip stretching across the sky. Ancient Greeks called it Γαλαξίας (Galaxias), which gave origin to the name "Galaxy." This word means "milky," because, in Greek mythology, it is the milk spurt from the breasts of Hera, Zeus's wife, while she was nursing the baby Heracles. The Romans inherited the Greeks' gods and mythology, and called this spectacular structure "Via Lactea," which means "Milky Way."

To the naked eye, it appears as a sort of nebula crossing the sky side to side, but when Galileo Galilei (1564–1642) pointed his spyglass at it, he realized, for the first time, that this nebula was made up of a huge number of very faint stars, something that had previously only be speculated upon.

This was just the first step toward a better understanding of the real nature of this object, and one of the first steps of Galileo's personal quest against the Ptolemaic view, which placed the Earth at the center of the universe.

The existence of such an immense number of stars in a specific strip of the sky, in fact, was hardly compatible with the classical model of a sphere of fixed, immutable and evenly distributed stars around a central Earth. If the galactic stars were to be of the same kind as the "normal" ones, the shape of the universe had to be completely different.

In the mid-18th century, Newton's theory had spread throughout the scientific community; its success in explaining the dynamics of the solar system was undeniable, and it was not limited to its planets, but also included objects, like comets, that had always teased human imagination. In this context, scientists like

© Springer Nature Switzerland AG 2019
G. Bernardi and A. Vecchiato, *Understanding Gaia*,
Springer Praxis Books, https://doi.org/10.1007/978-3-030-11449-7_2

Thomas Wright (1711–1786) and philosophers like Immanuel Kant (1724–1804) started reasoning that, given the universality of gravity, the Milky Way had to be an immense ring-like, or, better yet, disk-like, structure made of stars whirling around a center, like the planets around the Sun.

A few years later, William Herschel (1738–1822) started a pioneering project, thanks to his large telescopes, which were the most powerful instruments of their epoch. His goal was nothing less than creating a 3D chart of the entire Milky Way. The boldness of such an objective can be well understood by recalling that the first reliable estimation of a stellar distance was more than 50 years ahead, and only with Gaia, almost 250 years later, can we hope to more or less completely fulfill this ambition.

But if Herschel had no trustworthy method by which to measure stellar distances, how in the world could he dare to even start such a formidable endeavor? For the simple reason that he made the fundamental assumption that all of the stars had more or less the same intrinsic luminosity. It is known how the luminosity of an object decreases with distance, as we will see in a future chapter, so it was possible to estimate the distance ratios of all of the stars in the sky by comparing their apparent luminosities. And, since we know the distance of at least one star, the Sun, it was also possible to determine their actual distances.

This appears to be quite a strong assumption today, but we have to remember that, at his time, little or nothing at all was known of the nature of a star, and thus such a premise could be considered acceptable, at least as a first working hypothesis.

His patient work required the systematic observation of more than 600 regions of the sky, in each of which he took note (well before the advent of photographic astronomy, and, for that matter, of photography itself) of all of the stars accessible to his telescope and their apparent brightness. From the result of this first tentative charting experiment, our galaxy appeared to have quite an irregular shape, but the most surprising outcome was the position of just one of these stars.

Ironically, in fact, about one and a half centuries after the Copernican revolution had dislodged the Earth and humankind from the center of the universe, Herschel's results seemed to suggest that, after all, our planet was not so distant from this special place, because the Sun was more or less at the center of the Galaxy, which, at the time, could be rightly considered the whole universe.

The picture that we have of our galaxy today—let alone of our universe—is completely different. We know that the Milky Way counts hundreds of billions of stars, whose disk-like distribution is definitely not uniform; rather, it looks like a kind of central ball, called a "bulge," surrounded by a disk. The bulge is probably crossed by a bar, so that it more resembles a football than a soccer ball, and, viewed from the top, the stars of the disk are more crowded in some spiral-shaped regions that depart from the bulge, the spiral arms. From an edge-on view, instead, disk stars can be grouped together as belonging to either the thin or the thick disk, in which the latter is 3–10 times thicker than the former, and with a population that is generally older. The last structure identifiable in the Milky Way is the so-called stellar halo. It is typically formed by old stars grouped in tightly-packed "bunches" of some hundreds of thousands called "globular clusters," because of their approximately spherical shape. We know of about 150 of these clusters, which, in their turn, have an approximate spherical distribution around the galactic center.

We owe to these clusters the discovery that the Sun is actually quite far from the center of the Milky Way. So, how could Herschel have gotten it so wrong? And who fixed this flaw? And how did they do it?

In its bare bones, the answer to the first question is simply "dust." We have already mentioned that Herschel based his reconstruction on the wrong assumption that all of the stars had more or less the same luminosity. This hypothesis alone would have degraded its results, but the very reason that the Sun was eventually placed at the center of the galaxy is based on another assumption that Herschel was probably not even aware of: he was implicitly assuming that nothing could block his view of the stars. Again, this is completely wrong, because the Milky Way is not only made up of stars, but also of a huge amount of hot gas and an equally huge amount of dust, a term that astrophysicists use to indicate tiny solid particles that are typically of a size on the order of a micron (one millionth of a meter). Dust behaves like fog around the Earth: it scatters light away from its original direction, making stars redder and dimmer. We know what the effect of fog is; the farther away a light source is, the more its light is scattered. Instead of appearing as an object with a well-defined shape, its edges start to look blurred, and, after a certain distance, the source will be completely dispersed, reaching us from any direction, so that we will no longer perceive a single source, but rather an annoying glow coming from all of the foggy air. Thus, in this way, we can only see up to the same distance in any direction, and automatically place ourselves at the center of the sources that the fog allows us to see. Had Herschel known of the existence of the dust, maybe he would have reached the same conclusions, but he didn't.

This state of affairs changed dramatically when we became able to place the globular clusters around the galactic center, as we mentioned above, thanks to certain special stars, to the woman who discovered them, and to the man who sought them in these clusters.

The stars are called Cepheids, and, as we will explain in Chap. 10, they are a particular class of variable star whose luminosity varies according to a specific law that connects their period and absolute magnitude. This peculiar characteristic makes them a sort of stellar yardstick that allows for a precise determination of their distance. The relation between the period and luminosity of Cepheid stars was discovered by the American astronomer Henrietta Leavitt (1868–1921) in 1908, and, a few years later, another American astronomer, Harlow Shapley (1885–1972), sought out such stars in globular clusters. He was then able to determine their distance with enough precision to understand that these structures were distributed more or less on a sphere, but that its center was quite distant from the Sun. It was reasonable to suppose that such a center was that of the Milky Way.

Meanwhile, over a period of several decades crossing the 19th and 20th centuries, scientists managed to achieve a reasonable understanding of the composition and evolution of the "bricks" of our galactic building: the stars.

2.2   The Stars

The stars, like our Sun, are huge spheres of gas whose main components are hydrogen and helium, the two most common elements in the Universe. At its center, the gases are so hot and compressed that they burn, typically converting hydrogen into helium, but we know that hot gases tend to expand, so what prevents a star from dissolving and disappearing in space? The answer is "gravity."

This force among the star's particles pulls each of them towards the center, hindering expansion, and when one is exactly balanced by the other, the star appears as a spherical object whose center burns at a temperature of millions of degrees (Fig. 2.1). Their lives are very long with respect to the human scale, but they can vary a lot from star to star, from a few million to many billions of years. What makes the difference is mainly their initial mass; for example, stars twenty times more massive than the Sun will have a life roughly four hundred times shorter, but still counting in the millions of years.

Actually, gravity governs every phase of a star's life, beginning at its birth. The gas of which stars are formed is initially dispersed in space, and it is the gravitational interaction among its particles that gathers them. As long as they assemble together, the process accelerates, and temperatures start to increase at the center of the proto-stellar nebula. At first, this is just the natural heating due to the compression of the gas, but when temperatures are sufficiently high, at about 10 million degrees, the nebula literally "lights up," and a new star is born.

The ignition process is nothing else but thermonuclear fusion. As its very name says, nuclear fusion is a reaction in which two atomic nuclei get together, forming the nucleus of another atom of a different species. When this happens, the mass of the new nucleus is slightly less than the sum of the two original ones, and the difference is transformed into energy. The entire process thus transforms lighter atoms into heavier ones and releases energy.

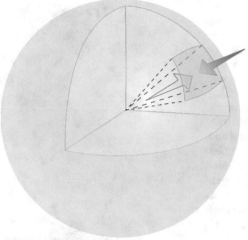

(c) Alberto Vecchiato and Gabriella Bernardi

Fig. 2.1 Stars are made of high-temperature gases that would be quickly dispersed throughout space without the attractive force of gravity. *Credits* Gabriella Bernardi and Alberto Vecchiato

Atomic nuclei are formed by protons and neutrons, and the former are positively charged particles, which we know repel each other because of their electromagnetic interaction, so how can such fusion happen? It is because a new force, called strong interaction, comes into play. Quantum physics tells us that this is an interaction among the particles that form the atomic nuclei, and that it can be much stronger than gravity or electromagnetic forces, but, at the same time, as distance increases, its magnitude decreases much faster than the previous ones. So, these attracting nuclear forces can vanquish the repulsive electromagnetic ones only when the nuclei are close enough. This is why nuclear fusion requires high temperatures of millions of degrees. In fact, to reduce the nuclear distance and cross the line at which the strong interaction wins out over the electromagnetic one, there is no other choice beyond setting them on such a violent collision course that the repulsion of both of the surrounding "shields" of electrons and protons can be overcome for an instant.

In this case, when a new star starts burning from a condensing nebula, the fuel of the thermonuclear reactions is the hydrogen. Two of these atoms merge together into one of helium, and, as we just mentioned, this process releases energy that further heats the gas up (Fig. 2.2). As we have seen, the hotter the gas, the more it generates a "negative pressure" that makes it expand, thus opposing the gravitational compression. If the negative expanding pressure is not enough to stop the positive compressing one due to gravity, the newly born star keeps collapsing in on itself. This again increases the temperatures, both because of the

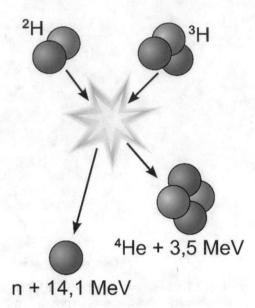

Fig. 2.2 Schema of the hydrogen-based thermonuclear reaction process, in which two isotopes of hydrogen (Deuterium and Tritium) collide, forming one Helium atom, in its ^{4}He isotopic variant with two protons, a neutron and energy. The most common reaction in the stellar interiors involves 6 hydrogen atoms that merge together, forming a ^{4}He isotope plus 2 protons, neutrinos and energy

^{2}H

^{3}H

^{4}He + 3,5 MeV

n + 14,1 MeV

increasing compression and because nuclear reactions are more efficient at higher temperature and pressure. When the temperature is too high, instead, and the negative pressure wins out over the positive one, the star expands, but, in this case, its density decreases and the temperature drops.

So, this is a self-regulating process that eventually puts the star in an equilibrium phase in which the rate of nuclear fusion is roughly constant, and just enough to prevent further progress of the gravitational collapse. This phase can last until there is enough "fuel" to maintain such a constant fusion rate. As long as hydrogen is burned into helium, in fact, the stellar nucleus will be "contaminated" by this byproduct of the nuclear fusion. This means that less hydrogen remains to be consumed, and also that it becomes more dispersed because it is gradually replaced by helium, dwindling the collision rate among hydrogen nuclei.

The actual duration of the equilibrium phase depends on the initial mass of the star. More massive stars, in fact, will need higher temperatures to stop gravitational collapse, and, as we have just seen, this implies a more efficient, and thus faster, depletion of the hydrogen reservoir. The Sun will take 5 billion years to run out of hydrogen, whereas a 10-solar mass star would take around 30 million years.

The life of a star as an object sustained and lit by thermonuclear fusion comes to an end when it can no longer burn its gas; the way in which the end comes, however, always depends on the initial mass, and, in general, the bigger the star, the more violent its "death" (Fig. 2.3).

In fact, hydrogen isn't the only atom that can be burned by nuclear fusion; other elements can do so as well, but, since they contain more charged particles, their repulsive force is higher, so they need higher temperatures to merge together. If

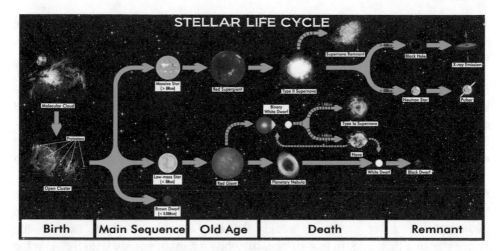

Fig. 2.3 Representation of the complete stellar life cycle according to the star's mass at the various stages (*Credits* Pelgrane, Wikimedia Commons)

hydrogen starts nuclear fusion at 10 million degrees, the next lighter element, helium, can burn only at 100 million degrees.

Since helium is heavier than hydrogen, it falls down to the center of the star, whose nucleus starts contracting again. The rising temperature of the helium nucleus may not be enough to ignite the fusion of this more massive element, but it releases energy by the good old compression mechanism, and this energy heats up the surrounding hydrogen. In a relatively thin shell, such temperatures can be high enough to restart the hydrogen nuclear fusion, so this shell expands, and the star suddenly "inflates" its external atmosphere, becoming an even larger, by a factor of 100, red giant, even though it is, at the same time, much less dense. What happens next, once again, depends on the stellar mass.

It might be that the mass of the helium nucleus is not large enough, so its gravitational contraction can be stopped by negative pressure of the temperature before the crucial 100-million-degree threshold. In this case, the inflated star simply disperses its external envelope, exposing the denser and hotter internal part and becoming a white dwarf that can only cool down gradually, like a burning brand. If, however, the mass is large enough, then the helium can ignite by nuclear fusion, and a new equilibrium phase sustained by this reaction can start.

Helium fusion produces an even heavier element, carbon, and the new nucleus evolves similarly to the previous one, but with these two elements in place of the former hydrogen and helium. Also, the amount of fuel is smaller than in the previous phase, so this new equilibrium can last for a shorter time.

This succession of shorter and shorter equilibrium phases, with a quick and temporary setting in periods in which the star burns like a kind of "cosmic onion," can go on for a while, because heavier and heavier elements start burning in the nucleus. This, obviously, is only if the star is massive enough to crunch and ignite these increasingly resilient atoms!

The nuclear fusion ladder climbs up the periodic table, passing the torch on to, in order, Carbon, Nitrogen, Oxygen, and so on, until it stops at iron; why? Simply because the nuclear fusion of two iron nuclei is endothermic, which means that it does not generate energy, but rather absorbs it!

At this point, therefore, there is no way left to sustain a burning star, and no nuclear fusion can stop its gravitational collapse. Moreover, the top-level reaction requires temperatures as high as 1 billion degrees, at which point atoms them-selves start to be crushed in the collisions. Protons and electrons can combine together, forming neutrons and other extremely tiny and fast particles called neutrinos. The latter interact so weakly that they can basically pass through the entire star in a jiffy, and, in this escape, they bring a lot of energy with them. What remains of the nucleus is colder, but, since neutrinos have a negligible mass, the gravitational pull remains almost the same, and there is even less energy left to stop it. The more the star collapses, the denser it becomes, and the more efficient the reaction is that converts protons and electrons into neutrons and neutrinos, which, in their turn, subtract additional energy from the nucleus. This ever-growing phase also used to be called the "URCA process," after the name of a famous casino in Rio de Janeiro, where the customers were induced to raise their bets on increasing losses. In this way, the entire atomic structure of the stellar nucleus can be converted into neutrons in a puff, and the star ends up in a big ball made almost entirely of neutrons. Well, maybe not so big, because neutrons, lacking any electromagnetic charge, can be compacted into a much smaller space than protons, so small that the entire mass of the Sun could be concentrated in a sphere of a few kilometers' radius.

Such an extreme situation creates the void between the nucleus and the remaining external parts of the star, which thus behave like a building without its basement; they crash at a fantastic speed towards the center, generating a shockwave that completely destroys the star, except for its neutron-made nucleus. We can see this shockwave as an incredibly bright, violent and fast explosion that astronomers call a supernova.

If the mass of the remaining stellar nucleus does not exceed a given limit, approximately between 1.5 and 3 solar masses, the game is more or less over. What remains of the once huge object is a few kilometer-wide neutron star, which spins at an incredible frequency. This happens because the original star was spinning at the beginning of the collapse, and it behaves like a classic dancer who starts twirling with open arms and then, suddenly, closes them in on herself.

If, however, the mass is still larger than this limit, the gravitational collapse is unstoppable, even for the neutrons. There are speculations as to the possibility that an even more extreme state of the matter formed by a quark's plasma can put this fall to an end, but no confirmation of this hypothesis yet exists. What we know for sure, instead, is that, at a certain point, the matter becomes so dense that nothing can escape its gravitational pull. Even photons are held back, so the resulting object transforms into an invisible, kilometer-sized stellar black hole.

In one way or another, regardless of the ultimate fate of the star, a great part of its former atoms are dispersed through space, but they certainly aren't wasted, because other stars and planets will form out of them.

In this way, for example, the oxygen that we breathe, the carbon in the cells of our bodies, and the calcium in our bones are all chemical elements that were produced inside of old stars billions of years ago.

Stars are different not only because of their many possible ends, but also in regard to their dimensions, colors, chemical compositions and luminosities. This information helps astronomers to understand a lot about their age and evolution.

2.3   The H-R Diagram

One very useful tool for identifying and using these different characteristics is the H-R diagram, or HRD, whose name comes from the scientists Ejnar Hertzsprung (1873–1967) and Henry Norris Russell (1877–1957). These astronomers, independently in 1905 and 1913, respectively, noticed that stars that have similar colors show some common features. In particular, they observed that blue stars are very bright, whereas red ones can be grouped into two sets, one bright and the other faint.

The color of an object is directly linked to the temperature of its surface; this is related to an ideal case studied in physics called a black body, which represents an object that can absorb every bit of energy hitting its surface, re-emitting it as electromagnetic energy of different frequencies. We will show the characteristics of the black body radiation curve in more detail in Chap. 7, and the mechanism can only be understood with quantum physics, but this analysis shows that such energy is distributed in the spectrum with a characteristic shape and specific features that vary as a function of the temperature of the body.

Referring to Fig. 7.6 of Chap. 7 for a graphical representation of the black body radiation curve, one can notice that, in all cases, the distribution starts at zero for infinite frequencies (zero wavelength), then grows until it reaches a maximum and decreases, tending to zero when frequencies go to zero (and wavelengths go to infinity). However, the higher the temperature, the more the curve peak is shifted towards high frequencies.

Human eyes perceive only a narrow band of electromagnetic frequencies as colors, blue being at the highest end, while red is at the lowest one, and the peak means that the corresponding frequency is the most prominent color—in case it falls in the visible band—of that distribution. So, a bluish body has a higher temperature than a reddish one. Actual bodies are different from perfect black bodies, but, qualitatively, their behavior is similar, and, in fact, embers change their color from bright red to dark red while they cool down. This is also reminiscent of the expression "white-hot" to indicate a very high temperature.

The color that we see is obviously that at the surface of the bodies, and we directly associate it with their surface temperature; the colder ones are red and the hotter ones are blue.

Hertzsprung and Russell placed the stars on a diagram with the temperature on the x-axis. For the above consideration, the temperature was that of the stellar photosphere, that is, the external visible shell of the star, computed as if they were perfect black bodies, which is a fair enough approximation. As in the black body plot, in the H-R diagram, the highest temperatures are on the left, while the lower ones on the right.

In the y-axis, there is a quantity that measures how much bright the star is, the absolute magnitude, or the absolute visual luminosity, that is, the amount of light emitted by the object.

This diagram shows that some combinations of luminosity and temperature occur much more often than others, and that stars seem to distribute mainly along two strips.

The first is a diagonal that cuts the diagram from the top left to the bottom right corner; the second, more or less horizontal, goes from the center to the far right. Approximately, this diagram looks like an overturned "seven" (Fig. 2.4). In the first group, we can find the so-called main sequence stars, whereas the second one contains the red giants or the supergiants. The main sequence is referred to in this way because the vast majority of the "normal" stars spend most of their lives here, and, as a consequence, most of them are on the namesake line of the diagram. The red giant line, contrastingly, is less populated, though these stars are generally brighter and thus observable at greater distances. In addition to these two, another one can be formed down on the left, smaller and less numerous. Here is where we find the white dwarfs mentioned above.

Astronomers say that the H-R diagram is a treasure-trove of information about stars, if one knows how to read it, so let's try and see if we can learn how to dig up some nuggets!

First of all, at the top of the diagram, in general, stars have larger radii, whereas at the bottom, they are smaller. This is because larger objects radiate more light, and thus are brighter, than smaller ones at the same temperature. For example, the diagram tells us that stars in the red giant branch are bright, but this can happen only if they are very big, compared to those with the same color/temperature in the main sequence. At the same time, they are more often on the right side, which

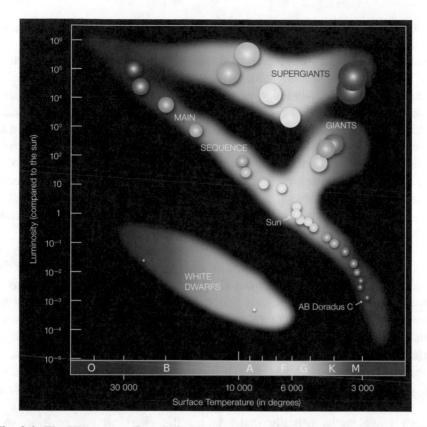

Fig. 2.4 The Hertzsprung-Russel diagram and the corresponding spectral types. (ESO Press Photo 28c/07, 19 June 2007)

means that their surface temperature is relatively low. The color of the white dwarfs, instead, suggests that their temperature is typically high, so they emit a lot of energy per unit surface, and this means that they must be small, because being placed at the bottom of the diagram is associated with a low total brightness.

In spite of their small size, white dwarfs also have a mass similar to that of the other stars, and thus their density is very high, much higher than that of a giant star. Indeed, these stars can reach a density of one ton per cubic centimeter, as if a little cube with a side of 3 cm weighed the same as a fully loaded truck.

The H-R diagram is so important to astronomy because the placement of a star within it depends on many factors other than those in the two simple cases shown above. Its mass, chemical composition, and age contribute to the determination of an instantaneous position, and from mass, chemical composition and a physical model of the stellar evolution, we can trace a specific evolutionary track of how the temperature and brightness of a star has varied over time. Seen from the other side, a precise determination of these two quantities can give us precious information on stellar masses, chemical composition and evolution.

This diagram is therefore a powerful tool for deducing important information about the stars, but, like any tool, it has to be built correctly. Ultimately, its accuracy depends on knowledge of the evolutionary models, so the better this knowledge is, the more precise the answers that the H-R diagram can provide. In their turn, these models can be constrained by populating the diagram with a large number of precisely positioned stars. Are we entering a vicious circle? Not really. It is just a matter of "bootstrapping our knowledge." What is needed is to characterize many stars by estimating, in an independent way, all of the necessary parameters (mass, brightness, age, etc.). These can then be used to build the diagram and to improve the models. From this point on, these models and the diagram become useful for other, less precisely known stars.

The peephole through which we can "observe" the chemical composition of the stars is the so-called spectral type, that is, the type of spectrum of the light that it emits. The light that we receive from an object, stars included, can be formed by several constituent wavelengths, of which we can perceive only a narrow range that we see as colors. When such light passes through a prism, or through another instrument called diffraction grating, it splits up into a parade of constituent wavelengths, called a spectrum, which can be recorded on appropriate detectors. As we have just seen, for a star, such a spectrum is approximately that of a black body, in which the intensity at each wavelength is determined by the temperature of the star. However, this is just a "background feature," and the intensity of each of these wavelengths actually depends on the chemical composition of the emitting object.

The light of stars is emitted by their external shell, or photosphere, so, from this spectrum, we can deduce their chemical constituents. This technique started to be applied in astronomy since the 19th century, and scientists soon realized that stellar spectra could be grouped according to specific attributes. Today, we use letters to identify each group; they are O, B, A, F, G, K, M, to which L, T and Y have recently been added. We owe this classification method to the American astronomer Annie Cannon (1863–1941). Her peculiar method of sorting has historical reasons. Previously, in fact, spectra were identified by the letters A to O, sorted alphabetically, and assigned by decreasing quantity of hydrogen revealed from their spectra. Cannon, however, showed that some of the letters were actually associated with the same spectral type, and could thus be dropped. She also made the important discovery that spectra could be associated with the temperature of the emitting photosphere, so she ordered the remaining letters by their decreasing values. Someone also invented the mnemonic sentence "Oh, Be A Fine Girl Kiss Me" to remember the original sequence more easily!

Because of the association between spectral types and temperatures, the former can replace the latter in the *x*-axis of the H-R diagram. This should sound familiar to our ears, because we know that a spectrum is nothing more than a "split up rendition" of the overall stellar color, and we have already mentioned that colors and temperatures are associated. Indeed, stars with spectral type O are blue, and they go to the red end as we skim over the sequence down to the letter M. Analyzing their spectra, then, is how astronomers estimate the temperature of a star; but there's more than this.

As our ability to recognize tiny features in a spectrum improved, each spectral type was subdivided into sub-classes identified by digits from 0 to 9. The lower the number, the higher the temperature, so that, for example, a star classified as A9 is closer to an F0 than to an A0.

Moreover, some peculiar features of the spectrum, called absorption lines, can be wider or thinner, and it was discovered that this width can be associated with the density of the photosphere. Previously, we have seen that a giant star has a very rarefied external shell, and it is intuitive to understand that a white dwarf, conversely, has a very dense atmosphere, because of its reduced size, and thus larger gravity at surface. This characteristic can help us to tell these two types of stars apart.

Actually, we can do better than this, for it is possible to recognize several of these types, identified by roman numerals from I to VII, associated with super-giants down to white dwarfs, respectively. So spectral types can also provide hints on the dimensions of the stars, and since dimensions are also linked to their luminosity, this further level of classification is called luminosity class.

The Gaia satellite is making a census of a huge number of astronomical objects, actually, more than one billion, up to a luminosity more than 10 billion times dimmer than that of Sirius. Its goal is to measure the position and velocity of these stars, but also their luminosity, temperature and chemical composition. In addition to the stars of our galaxy, Gaia can target other kinds of object; for example, the Andromeda galaxy, the Magellanic clouds, or other galaxies, but also, at the opposite end in terms of cosmic size, planets and asteroids. But what is a galaxy? Or a planet? Let us start with the smaller ones.

2.4   Planets and Minor Bodies

We know that our solar system is made up of a large number of bodies, all of which are much smaller than the Sun and orbiting around it. These objects can belong to several classes; first, there are the planets, the larger ones, like Jupiter and Saturn, and the smaller ones, like the Earth and Venus.

The former, which also include Uranus and Neptune, are made mainly of hydrogen and helium, and form the set of gas giants, whereas the latter, along with Mercury and Mars, are the so-called rocky planets, because their composition is mainly formed by rocks and metals, with minimal quantities of these gases, so that, externally, they appear solid. These eight bodies can be surrounded by smaller copies of the rocky planets, the planetary satellites.

Dwarf planets can be found between Mars and Jupiter and beyond the orbit of Neptune. According to the 2006 definition of the International Astronomic Union, they are spherical planetary-sized bodies that have not been able to clear their orbits and form a proper planet, which means that we can find many similar bodies in similar orbits. This is the case of Ceres in the main asteroid belt beyond the red planet, and of Pluto and others after Neptune, in the region of the so-called Trans-Neptunian objects.

Asteroids, on the other hand, are even smaller and more or less irregularly-shaped bodies that can generally be found in large families, located in different regions of the solar system, like the already-cited main belt, or in orbits crossing those of the inner planets, or even around special points that trail the orbit of Jupiter. Finally, we come to the comets, sort of giant snowballs that come from the most remote boundaries of the solar system.

So, we have a large variety of non-stellar objects, and the study of their structure and evolution is often a multidisciplinary field of research that mixes astronomy, geophysics and geology. One reason that, in addition to this interaction with different disciplines, planetary science is sometimes divided from the other fields of astronomy is that we have only one example of a planetary system at hand, since, until recently, no planets were known except for those orbiting around the Sun.

There has been extreme interest in problems involving the formation and evolution of planetary systems, and many theories have been formulated in this regard since the 18th century. Is there a reason why the closest planets are also the rocky ones, whereas all of the giant planets orbit beyond the asteroids? Did the planets form together with the Sun, or maybe later? And, again, what happens to the planets when a star like the Sun becomes a red giant?

There are reasonable answers to many of these problems, for example, it is clear that giant planets can exist only at a certain distance from a star, because the solar wind clears the inner regions from the light elements that constitute the largest mass fraction of these objects, but, without other cases to study, no real progress could be made in this regard.

At the end of the previous century, the situation started to change radically. As we will see in more detail in a later chapter, after the discovery of the first exoplanet, that is, of a planet around another star, this field of research flourished, and several thousands of these objects have been discovered to the present day.

The development of this new astronomical discipline, however, has contributed more to the creation of new questions, rather than to answering the old ones. For example, gas giants have been discovered orbiting very close to their parent stars; if they can exist only in relatively large orbits, how did they arrive there? This sort of question, and many others, now waits for a solution, but perhaps the most popular one is about alien life.

Before we started to get to know our solar system better, it was generally believed that life was quite common on other planets, while, nowadays, the general understanding is that, although recent researches suggest that there are still other possible candidates for hosting life in the solar system, it might be that life exists only on Earth. But if this is true for our system, what about those of other stars? Once again, our limited knowledge makes it very hard to give a reliable answer, and the best example of our ignorance is the Drake equation.

In the early' 60s, the astronomer and astrophysicist Frank Drake, in an attempt to reckon all of the factors that had to be known in order to establish a program for the search for intelligent extraterrestrial life, elaborated the equation bearing his name. The number N of planets that can host intelligent lifeforms, in the sense that

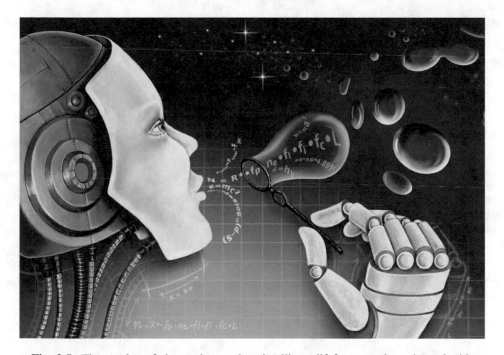

Fig. 2.5 The number of planets that can host intelligent lifeforms can be estimated with the Drake Equation, illustrated here along with other famous ones. At present, however, our knowledge is so limited that the possible answers can vary in the range between one (us) to many millions (*Credits* Danielle Futselaar)

we can communicate with them, is $N = R* \cdot f_p \cdot n_e \cdot f_l \cdot f_i \cdot f_c \cdot L$ (Fig. 2.5), according to the factors identified by Drake. For example, $R*$ represents the average rate of star formation in our galaxy, that is, the frequency with which new stars are formed. In this case, we have a good estimate for this value (about 7 per year), but in other cases, like that of f_i—the fraction of planets hosting life that evolves in intelligent forms—the actual value is completely unknown. One might presume that, given enough time, sooner or later, life always evolve into intelligent forms, but we have no reason to prefer this hypothesis to the opposite one, and think that it is very unlikely that life evolves into intelligent forms. In the first case, f_i would be 1, in the other one, it would be a very small number, which gives two completely different scenarios.

Fiddling with these numbers is extremely easy, and one can come to the conclusion that the Milky Way currently hosts many millions of intelligent life-forms eager to communicate with humans or that we are alone in our galaxy with equal certainty!

3

Galaxies Everywhere!

3.1 From "Galaxy" to "Galaxies"

As we have seen, William Herschel's first pioneering attempt at mapping the Milky Way resulted in a very irregular structure, with the Sun, and thus the Earth, approximately at its center. This new "Ptolemaic" vision of the universe was finally proven wrong by the work of Shapley, who showed, at the beginning of the 20th century, that the spherical distribution of the globular clusters located the gravitational center of the Galaxy far away from the Sun, once again demoting our place to its "Copernican-aware" non-special role.

However, the dispute between the Ptolemaic and Copernican concepts of the universe seems to recur every time scientists discover that the cosmic boundaries are larger than that which was previously thought. Soon after Shapley's discovery, another astronomer started to cast doubts on the validity of the equation "Milky Way = Universe" that was implicit in the understanding of the time.

His name was Heber Curtis, and he claimed that some well-known objects, like Andromeda, were not just nebulae inside our galaxy, but rather extragalactic objects. Actually, he argued that they were other galaxies on their own!

Since the apparent size of these objects in the sky was tiny compared to that of the Milky Way, this implied that intergalactic distances were enormous compared to the dimensions of our galaxy, and thus that the size of the universe had to be extended by a factor of at least thousands. This was hardly believable for many astronomers, and Shapley was seemingly so vexed by this hypothesis that he took the flag of the opposing team and started what is now called "The Great Debate" of astronomy.

© Springer Nature Switzerland AG 2019
G. Bernardi and A. Vecchiato, *Understanding Gaia*,
Springer Praxis Books, https://doi.org/10.1007/978-3-030-11449-7_3

It officially began in 1920, with a double conference by Shapley and Curtis at the National Museum of Natural History, Washington D.C., but it did not come to an end until some years later.

Actually, Curtis was not a lonesome in his views, for he had embraced a hypothesis already formulated in mid-1700s and supported by Immanuel Kant, known as that of the "island universes," and there was no lack of interesting clues in its favor. For example, the American astronomer stated that the amount of novae (stellar explosions similar in their aspect to the supernovae, but less powerful) was larger in the direction of the Andromeda nebula (Fig. 3.1) than in other directions in the sky, and that their luminosities were systematically dimmer. From these, he could derive an estimation of the distance of this object that, although quite inaccurate, placed it well outside of the boundaries of the Milky Way. On the other side, Curtis' statement had definitely bold implications; not only did it require a complete revision about the dimensions of the universe, but one would have to admit the existence of objects of inconceivable luminosity, known today as supernovae.

Fig. 3.1 The Andromeda galaxy is among the closest galactic neighbors of the Milky Way. It lies about 2.5 million light years from us, and it is the only one visible to the naked eye in the northern hemisphere (*Credits* Francesco Meschia)

In 1923, however, another astronomer, who would later give birth to modern cosmology, made a discovery that put a first but decisive nail in the coffin of the idea that the Milky Way was the only galaxy in the universe, and certainly ended the notion that it was actually the entire universe itself. His name was Edwin Hubble, and the discovery was that of a Cepheid in the direction of the Andromeda galaxy, the resulting verdict from which was unambiguous: the distance of the Andromeda Cepheid was almost ten times greater than the diameter of our Galaxy. Now, one could insist that Andromeda was still a galactic nebula, and that its correspondence with the Cepheid was just a matter of perspective, but this would have made little difference, since at least one star was certainly beyond any previously accepted distance. Curiously, the same type of stars that helped Shapley to make his greatest discovery was also decisive in disproving his strongest conviction.

We started this section by anticipating that the changing of the scale of the universe established after the Great Debate started another round of the old Ptolemaic versus Copernican match; yet, up to now, it would seem that we have simply moved from one Copernican perspective to an even more extreme version of it. We will soon see that this is not the end of the story, but, before the final "coup de theatre," we need to know more about these new islands of the universe: the galaxies.

3.2   Galaxies Have a Dark Side. Or not?

Thanks also to Hubble's findings, we know that a galaxy is a huge ensemble of stars, gas, and interstellar nebulae made of dust. A typical galaxy, like the Milky Way, counts some hundreds of billion stars, and gas and dust is distributed more or less evenly among them. We also know that there are hundreds of billions of galaxies in the universe accessible to our instruments, and that they can take different forms, but the way that they formed and how they evolved still represents an unsolved problem for astronomers, or at least a problem with, literally, many dark sides, as we will show in a few paragraphs.

Edwin Hubble is again one of the main characters of this scientific endeavor. In the 1920s and 1930s, the American astronomer worked at Mount Wilson observatory, California, and was able to use its recently completed telescope, the largest and most powerful of the time, to study galaxies. In particular, he measured their spectra and studied their morphology, that is, their shapes, as a first classification attempt to help to understand the origin and evolution of these huge "bricks" of the universe.

Hubble's effort produced the so-called "tuning-fork diagram," because, in his morphologic classification, the way that the shapes of the galaxies are arranged resembles that of the namesake device used for tuning musical instruments. In this diagram, galaxies are divided into three main categories: those with circular or elliptical shape are placed on the grip, while those sporting spirals arms with or without a central bar sit on each of the two prongs, respectively.

To sort out each kind of shape, Hubble adopted a classification schema similar to that of stellar spectra; Capital letters are used for each type of shape, so E is used for elliptical galaxies, S for spirals and SB for barred spirals, while variations within each type are identified with numbers or lowercase letters (Fig. 3.2). In more recent times, this diagram underwent several modifications so as to take into account a more detailed shape differentiation, as well as other characteristics, like spectral type or brightness. For example, both the Milky Way and the beautiful M31 or Andromeda galaxy, the only one visible to the naked eye from the northern hemisphere, belong to the subclass Sb, but more recent investigations suggest that, instead, our galaxy might have a bar, and thus should be classified as SBb. Another system called Hubble-de Vaucouleurs makes a further distinction between purely spiral galaxies (SA), purely barred (SB) and intermediate or weakly barred (SAB), which gives its diagram the appearance of a three-pronged fork, and, in this system, the Milky Way might be attributed the SAB class. There also exist galaxies that do not belong to any of these types and are classified as irregular (I) because it is not possible to attribute any specific shape to them. The Large Magellanic Cloud and the Small Magellanic Cloud, visible to the naked eye from the southern hemisphere, are two examples of this type.

Spiral and irregular galaxies are rich with gas and dust, and, in the arms of the spirals, one can usually observe blue-colored and young stars, whereas the

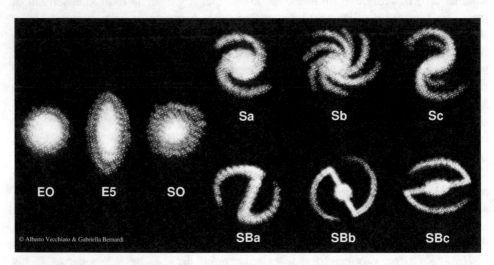

Fig. 3.2 Schematic view of the Hubble classification schema for the galaxies (*Credits* Gabriella Bernardi and Alberto Vecchiato)

elliptical ones usually have a very brilliant central region, with little gas and dust, and their stars are mostly red and metal-poor—an expression indicating that they have a particularly small fraction of elements other than hydrogen and helium. The characteristics of the elliptical galaxies are associated with old stars, which suggests that they are generally more ancient and evolved than spiral ones. This is why, initially, an evolutionary meaning was also attributed to the classification diagrams, in the sense that a galaxy started as an irregular ensemble of gas and dust that aggregates into stars and takes a spiral shape, to evolve finally into an elliptical one.

This sequence is useful for describing galaxies, and, indeed, it is still used to this aim, but its alleged connection with an evolutionary schema has been revealed to be inconsistent. Today, we know that the evolution of a galaxy is a process far more complex than was imagined in the past, depending not only on the initial composition, but also on events, like collisions with other objects, that may happen during its history, altering the galaxy's shape and composition.

Tracing back the history of a galaxy is an extremely complex task, which we are only beginning to understand. Scientists use very powerful supercomputers to simulate what could happen from some initial situation, but even supercomputers are not powerful enough to follow the motion of each single star, and a series of tricks and approximations has to be exploited to deal with this issue, which is made even more difficult by the need to treat other components, like gas and dust.

Another problem comes from the uncertainties that we still have about the behavior of gravity at large scales. The law of gravity presented by Newton in the second half of the 17th century is tightly linked to the three laws of planetary motion that Kepler had discovered about fifty years earlier, from the astronomical observations of planetary motion made by Tycho Brahe. These laws predict that the farther a planet is from the Sun, the slower its orbital velocity. Similar considerations can be made for the stars, thus deducing that their rotation speed around the center of a galaxy should decrease following a precise curve—similar to that of Kepler's laws when most of the mass lies inside of their orbits—that is a function of their distance from the central bulge. It is puzzling, then, to realize that such stars exhibit completely different behavior, and that the rotation curves remain almost constant for a long stretch, as if there was a large quantity of invisible mass beyond the visible boundaries of these cosmic islands.

It is necessary to be clearer about the actual meaning of invisible mass. Indeed, in space, we cannot see everything with our eyes. Light coming from astronomical sources is an electromagnetic signal that can reach our planet at a very wide range of frequencies, from radio to gamma rays, and our eyes can detect them only within a narrow range, which is thus called the visible band (Fig. 3.3). Our instruments, however, can be designed to detect any frequency of the

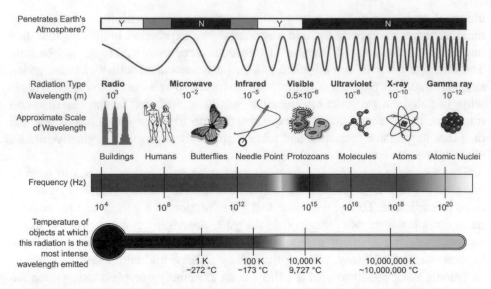

Fig. 3.3 Classification of electromagnetic waves according to their frequencies, and associated with various significant properties

electromagnetic spectrum, which gives us access to a large number of phenomena that would otherwise remain invisible, in the sense of "undetectable."

For example, there are peculiar galaxies that are called "active" because they are characterized by emissions of a large quantity of energy from their nucleus at frequencies outside of the range of the visible band. These kinds of object can be investigated with observations in the radio or X band, and such observations were begun in the 1970s using satellites equipped with particular detectors, because the atmosphere blocks the X-rays.

But, however different this light may be, the messenger is always the same kind of electromagnetic wave. The problem of the invisible mass, instead, is of a completely different kind. No observation in any band was able to detect the presence of additional masses emitting an electromagnetic signal, but the message from the laws of gravity could not be misinterpreted, and it was speaking loud and clear in favor of an additional large source of gravity, that is, of a large quantity of unseen mass.

Since this is the general behavior of any galaxy, the most accepted explanation is that which holds that such an invisible mass does indeed stretch beyond the boundaries of the galaxies, and astronomers call this invisible entity dark matter. Actually, it turned out that such a name was more befitting than one would even have initially imagined. Soon, it was discovered that such matter can be dark, but it weighs much more than the visible kind; to be precise, there is about five times more dark matter than all that we can reckon by summing up all of the stars and the gas and dust nebulae. And, as if that wasn't enough, as investigations continued, dark matter became more and more elusive. After more than 80 years, despite the incredible amount of theoretical and experimental effort carried out by astronomers and physicists, we still don't know what dark matter is made of. The key point here is that, in order to be detected, a particle must be able to interact with our detectors, and it seems that a fundamental property of dark matter is that it can interact in an incredibly feeble way, if in any way at all. One example of ordinary matter that behaves in a similar way is the neutrinos. They are tiny, uncountable particles, so many in number that trillions of neutrinos pass through our bodies every second, but we never notice them, because they interact so weakly that they can pass through the Earth almost undisturbed. But neutrinos cannot be the dark matter for two reasons; first, we could estimate their total mass despite all of the difficulties, and it is not enough to make up the difference, and second, because they are extremely fast. Dark matter cannot be fast; otherwise, the universe would look very differently from the way in which we see it. Since temperature is associated with the velocity of the particles in such a way that the higher the temperature, the higher the velocity, astronomers usually dub the two scenarios of fast neutrinos and unknown slow particles as "hot dark matter" and "cold dark matter," and experimental evidence supports the latter.

So, up to now, dark matter has still been undetectable, except for its gravitational signature. We have plenty of clues as to its existence, but they are practically only of the gravitational kind. For this reason, some scientists believe that it might be gravity that, under certain conditions, does not behave in the way that we think. Also, in this case, several efforts were undertaken to devise a reliable explanation, but results are still inconclusive.

All these kinds of problem, from the computational ones of our supercomputers to those with a more theoretical and experimental nature, are in desperate need of better data, which Gaia may provide with its precise catalog.

A similar problem emerges when we don't observe each galaxy as a single object, but rather focus solely on their interactions.

Telescopic surveys show that galaxies are more or less evenly distributed in space. However, as stars have the habit of grouping themselves in clusters or, at larger scales, in galaxies, so the latter tend to group themselves in groups of hundreds: the galaxy clusters (Fig. 3.4). They are the largest and most massive individual physical objects in the universe.

Their shape can be regular or not, but, in general, since their distribution is steered by gravitational force, galaxies in clusters are not scattered randomly. In a planet or a star, the heaviest elements tend to concentrate in the nucleus, and, similarly, the more massive galaxies concentrate in the central part of the cluster. In a very broad sense, all clusters should tend to assume a regular form by gravitational force that operates between other galaxies.

Fig. 3.4 Picture of the galaxy cluster LCDCS 0829 obtained by the Hubble space telescope (*Credits* ESA/Hubble & NASA)

We can measure the velocities of single galaxies so as to make deductions on the evolution of the cluster that they form. For example, we know that a velocity limit exists, called the escape velocity, above which a body can get free from the gravitational pull of another object; the larger the mass, the greater the velocity. For the same reason, by observing the motion of the galaxies in a cluster, we can estimate the mass that it must have to keep them tied together. As you might suspect, the velocities that have been measured are usually much bigger than the escape velocity required by the sum of the visible mass of all of the galaxies in the cluster. Obviously, it might be that, sometimes, a cluster appears just as it is because the individual galaxies did not have enough time to run away, but it is impossible to accept such an explanation for all of the clusters. Once again, it looks like another manifestation of the mysterious dark matter.

3.3 Are We at the Center of a Ripping Universe?

It is now time to explain why the discovery that the size of the universe extends far beyond that of the Milky Way again proposed the recurring dilemma of what is at the center of the universe.

Once again, Hubble is the "culprit," because, in 1929 and then again in 1931, in collaboration with Milton Humason, he proposed a law that unleashed a great deal of astonishment within the scientific community. Measuring their spectra, he could estimate the relative velocities of the galaxies with respect to us, and, using the Cepheids in these galaxies, he also obtained an estimation of their distances.

The result was that the velocities and distances were approximately proportional. In other words, the farther away the galaxy, the faster it receded from us. Normally, in a situation in which galaxies were more or less uniformly scattered around and were moving at random velocities, one would expect that some would be receding and others approaching, in no particular order. These data, instead, seemed to suggest that the galaxies in the universe were expanding, sort of like the debris from a giant explosion, and that we were at the center of such an explosion! Shapley had shown that we are far away from the center of the Milky Way, but he believed that the latter was the whole universe. Hubble, in his turn, proved that the universe is much larger, but he apparently also put us, or at least our galaxy, at its center, once again!

Fortunately, this cosmological dilemma had been already solved, even before its discovery by the Russian scientist Alexander Friedman and, later, in 1927, by the Belgian priest and physicist Georges Lemaitre. They had independently calculated that, by applying the laws of general relativity—the new theory of gravity proposed by Albert Einstein in 1915—to a simple but reasonable model of the universe, an expansion of the type later observed by Hubble had to occur. Such expansion then led to the theory known today as the Big Bang, but it was no explosion of the sort that we are used to. Rather, it was the 3D version of the inflating of a balloon observed from its surface.

If we draw some points on this surface and we start inflating the balloon in a way that its radius increases at a constant rate, then one point would see all of the others receding, following a Hubble-like law; the farther the points, the faster the receding speed. But this is not a peculiar trait of a single point; any point would observe the same thing. And since, in this 2D scenario, we are forced to stay on the surface, there is no such thing as the center of this expansion, because there is no center of the spherical surface.

It is difficult to understand this phenomenon from our 3D point of view, because expansion is not going on "immersed" in a 4D space, but this is what relativistic equations tell us. It is the space itself that is expanding like the surface of the 2D balloon, and the effect of such an expansion is the Hubble law.

Until about 20 years ago, this so-called cosmological model was more or less accepted by everybody in the scientific community. Among other things, it foresaw that the expansion rate would have to decrease over time. The debate focused on the possibility that there was not enough mass in the universe to invert the expansion. If so, at a certain point, the universe would have started contracting, probably returning to the state that originated the Big Bang; otherwise, it would have continued its expansion forever, although at slowing rates.

In 1998, Saul Perlmutter, Brian Schmidt and Adam Riess presented the results of their project, which had the goal of charting the expansion of the universe to the farthest possible limit by using the explosions of supernovae as distance indicators.

According to their measurements, the expansion rate of the universe did not decrease with time, but was rather accelerating!

These results simply subverted every cosmological model in existence, and scientists desperately tried to seek a reasonable explanation. The current, most accepted explanation is offered in the form of another dark side of the universe. But this time, the answer is even stranger than an unknown form of matter.

As weird it can be, dark matter follows the known rules of gravity, that is, it attracts normal and other dark matter. This new entity, instead, would have to explain a behavior that opposes gravity, because we have seen that, with or without its dark counterpart, matter in the universe should fight against expansion.

Observational data, instead, apparently tell us that, while the universe enlarges, an unknown force that is gravitationally repulsive comes into play, whose effect is to accelerate its expansion. Actually, Einstein had introduced a quantity in his equations called the cosmological constant, which had an exactly similar repulsive effect, but it became untenable after the discovery of the expansion of the universe, and the famous scientist himself dropped it as his "biggest blunder." The phenomenon of the accelerated expansion revived the idea under a different form called "dark energy," but, at present, we know practically nothing about this entity, except that, according to our measurements, the recipe for our universe is the following: 5% of ordinary matter, 27% of dark matter, and 68% of dark energy.

Not only is the nature of dark energy unknown, but the possible evolution of the universe has also unleashed the wildest of imaginations. For example, it was also supposed that, giving credit to the nature of this accelerated expansion and extrapolating its extent to its apparently natural consequences, there will come a time in the future when the expansion will not only move galaxies away from other galaxies, but the stars inside of the galaxies will also be separated, and then the planetary system, down to the smallest scale of matter, ripping off the very fabric of space-time.

Scientists are proposing a picture in which we do not truly understand what is going on, and we still need to accept that, at 95%, we have no idea of what is making this happen. Definitely, "it's a long way to Tipperary" for science and astronomy!

Part II
A as in Astronomy and Astrometry

But what sweet delirium is theirs, who fabricate
worlds without end, measuring with thumb
and thread the sun, moon, stars, spheres...
Erasmus of Rotterdam, In Praise of Folly, 1511

4

History of Astronomy at a Glance

4.1 Astronomy and Astrometry

In the two previous chapters, we learned about the astronomical targets of the measurements of Gaia. We also mentioned some of the scientific mysteries that scientists hope to solve with the help of this satellite, and stressed their importance for Astronomy and Astrophysics.

In the introduction, instead, we anticipated that this mission, with all of its advanced technology and ambitious goals, is the last heir of a discipline whose origin is lost in the mists of time, born when the boundaries between science and magic were still undefined. We referred to it by its modern name, Astrometry, but we could also have used the word Astronomy; however, we are introducing a lot of these "Astro-" words without explaining explicitly what they are all about. For example, we could say that, since, in ancient times, it was all about measuring the positions and the motions of the objects in the sky, it makes little sense to distinguish between Astronomy and Astrometry when we refer to remote historical periods, but it is better to put things in due order, and start from the very beginning.

The word astronomy comes from the Greek ἀστρονομία, which is a word composed of two roots, ἄστρον ("astron"), which means "the stars," and νόμος ("nomos") or "law." So, astronomy is the science that studies the laws of the stars, or, in general, of celestial phenomena.

Several archeological relics provide unquestionable witness that humans have been observing and using the sky and its phenomena since the Paleolithic Era. For example, one of the most ancient relics is a lunar calendar, represented through the phases of our satellite, carved in an animal bone dated to about 34,000 years ago, when the last Neanderthals were still living in Europe, during the latest glaciation

© Springer Nature Switzerland AG 2019
G. Bernardi and A. Vecchiato, *Understanding Gaia*,
Springer Praxis Books, https://doi.org/10.1007/978-3-030-11449-7_4

Fig. 4.1 The northern hemisphere reproduced in a table of the "Uranographia" by Johannes Hevelius (1687)

period. Other studies seem to suggest that the idea of grouping stars in constellations can also be dated back to prehistoric times.

In antiquity, then, astronomy was related to the mapping of stars in the sky, preparing calendars, predicting the motion of the "wandering stars," that is, the planets, and this sort of thing (Fig. 4.1). Nowadays, all of this is the subject of a sub-field of the overall astronomical science called Astrometry, which literally means "measure of the stars." So, astronomy evolved and expanded the subject of its studies progressively, and, eventually, its original realm became a part of a more complex picture. But how did it happen?

4.2 Astrometry and Celestial Mechanics

We do not want to underestimate the importance of non-Western astronomy, or of the researches conducted in epochs different from that of the so-called "scientific revolution," but this sketchy overview of the history of astronomy is tailored to explain the way in which what we currently call astronomy started to diversify in many different subfields, a process that began in the 17th century, with developments in the field of Mechanics, of the theory of gravitation, and of Calculus, started by Galileo, Newton, Leibniz, and others, and, for this reason, we have to concentrate on a very short historical period and on Western culture.

Roughly speaking, we can say that the first step of this scientific trail starts when attention is paid to the "why" part of the problem, instead of the "how." Knowing that Polaris remains fixed in the sky, or that when Sirius rises just minutes before the Sun (the so-called "heliacal rise"), the Nile is about to flood the fields, is surely useful for pointing yourself in the right direction or cultivating a good harvest, but astronomy as we know it today started to be shaped when Mechanics, Gravity and Calculus nurtured the development of a new discipline called Celestial Mechanics, which, in practice, applies the newly developed physical and mathematical theories to the problem of the motion of celestial objects.

These new studies completely revived the theoretical branch of astronomy, much like the introduction of the telescope revolutionized its observational part. Similarly to what happened in more recent times with Relativity and Quantum Mechanics, the impact of Celestial Mechanics transcended the boundaries of science, influencing the philosophical thought and, ultimately, the cultural mindset of the society of the 19th century.

The power of this new approach in determination of the motion of the celestial bodies was so impressive that many scientists and philosophers imagined the universe as a sort of huge clockwork ruled by the laws of nature, and, in particular, by gravity, whose evolution could be determined, in principle, if the initial positions and velocities of all of its elementary constituents were known. This philosophical tenet is called "determinism," and was a fundamental part of positivism, the dominant philosophical movement between the 19th and 20th centuries.

When Napoleon was presented with Laplace's book *Exposition du système du monde*, so the story goes, he provocatively complained to the author that such a huge opera about the universe did not even mention God, to which the scientist swiftly replied, "Sire, I had no need of that hypothesis."

Although apocryphal, this anecdote greatly illustrates the impact that we have just mentioned. Pierre-Simon de Laplace (1749–1827) was one of the first and most influential representatives of determinism, and we owe a significant debt to his monumental five volume work, containing the first systematization of Celestial Mechanics, of which the "Système du monde" was actually a non-mathematical exposition.

At this point, thus, we can consider Astronomy as a two-branched discipline, one theoretical and one observational, both dealing with the position and motion of celestial bodies, in which, roughly speaking, Celestial Mechanics corresponds to the former and Astrometry to the latter. A theoretical approach had existed since a very long time in the past, but the novelty now was the transition from the purely geometrical principles and mathematical techniques inherited from antiquity (mainly from the Greek and Hellenistic civilization) to the new ones, based on Newton's Physics and differential calculus.

4.3   Astrometry and Astrophysics

More or less in the same period, another evolution was occurring in Astronomy, an even more profound development that, eventually, completely reshaped this ancient science. At the beginning of the 19th century, it was very well known that light from a source could be separated into its color components, its spectrum, by special tools like prisms.

In the first half of the century, two physicists, Wollaston and Fraunhofer, independently examined the light from the Sun with decomposing devices, discovering that the spectrum was scattered by a huge number of dark lines. Some decades later, it was understood that the spectrum of light emitted by heated elements was arranged into separate lines, and the dark ones observed on the Sun corresponded to those emitted by elements like Hydrogen, Lithium, and others. For the first time, it was realized that Astronomy could use the light from the sources to study not only their motion, but also their composition, and that the stars are formed by the same material that can be found on the Earth. These events triggered an increasingly important involvement of other branches of Physics in Astronomy that, by the end of the 19th century, gave birth to a new discipline called Astrophysics. Its aim was to study the nature and physical evolution of celestial bodies, and allowed us, for example, to understand how stars are born, evolve and die.

The dawn of the 19th century also witnessed the important discovery that there existed other "kinds" of light, not visible to the human eye. Infrared frequencies were discovered by William Herschel in 1800, and, just one year later, it was the turn of ultraviolet. With appropriate devices, these parts of the electromagnetic spectrum could be detected and measured. The same happened in the 20th century, with radio, X and gamma frequencies. We realized that celestial objects could broadcast their signals using several types of messenger, which gave origin to even more types of astronomy and astrophysics subjects.

It was only in 20th century, however, that, for the first time, astronomers used a medium other than some kind of electromagnetic signal for their investigation. Neutrino astronomy was born in the late 1960s, when the Homestake Experiment detected the first solar neutrinos. We have already mentioned these particles, generated by particular radioactive decays or nuclear reactions, in a previous chapter, including the rather elusive character that makes them extremely difficult to detect. On the other hand, it is exactly this "reluctance" to interact that makes them so precious in astronomy, for they can carry information, for example, from out of the interior of the Sun, or from disruptive events like supernovae. These regions, in fact, are otherwise inaccessible by photons, because the interaction of these particles in such turbulent environments completely destroys any information about the interior.

The latest development in this flourishing of "multi-messenger astronomies" is still in its early days. In 2015, the first direct detection of gravitational waves, predicted by Einstein as a consequence of his General Relativity theory one hundred years earlier, opened the route to a "gravitational-wave astronomy" that can be used to explore different and yet inaccessible phenomena in our universe.

We now know more about the birth of Astronomy and of its expansion and evolution into a wide range of different branches; we learned the meaning and the scope of these sub-fields, and why it was necessary to invent a new word, Astrometry, for an ancient science that deals with the measuring of the position and motion of the objects in the sky. This discipline is the foundation for anything we would wish to know about Gaia, so it is time to better understand its history and how it works.

5

Astrometry

5.1   The What and the Why

So, Astrometry "just" deals with the mapping of the sky, whereas the "really interesting things" are investigated by the other, newer, and shinier branches that make up astronomy. But… is it really that simple?

Actually, it is certainly true that this discipline includes all of the techniques that allow us to measure precisely the so-called astrometric parameters of celestial objects, which means those describing their position and motion. This apparently limited task, however, is not simple at all, and, above all, it is at the very basis of astronomy, so that practically every sector of this ancient science depends on it. Only good astrometric data, as we will learn later, can produce meaningful science.

We need to discover more about astrometry to fully appreciate this bold statement, so let us start precisely from an understanding of what those above-mentioned astrometric parameters actually are.

5.2   Positions

We are accustomed to the fact that a position in our three-dimensional space is defined by three coordinates, but an astronomer, and, even more so, an astrometrist, would say instead that the word "position" refers to just two coordinates. This is not because they are thriftier with coordinates; rather, it is just a matter of convenience.

If the expression "celestial sphere" is a synonym for the sky, it is because, to our eyes, it looks like a spherical surface; we are not able to tell the distance of

© Springer Nature Switzerland AG 2019
G. Bernardi and A. Vecchiato, *Understanding Gaia*,
Springer Praxis Books, https://doi.org/10.1007/978-3-030-11449-7_5

objects in the sky. So, if we agree upon the use of polar coordinates, that is, longitude, latitude and distance, we only need the first two. Celestial longitudes and latitudes are used in the same way as they are on Earth, while distances, as we will see later, are treated differently.

Indeed, as longitudes and latitudes are expressed as angles, astrometry is all about angular measurement, and when, in a starry night, we observe, for example, that Polaris is less than 1° from the north celestial pole, this is already some sort of astrometry. A quite inaccurate one by today's standards, to tell the truth.

Simply recording the positions of stars and of objects in the sky in general can look simple, in fact, but everything depends on the precision that we want to achieve.

The human eye has an angular resolution of approximately 1 arcmin, which means that this is the limit at which the average man or woman is capable of recognizing two separate objects; at smaller (angular) distances, they will be seen as a single object. But what exactly is an arcminute?

First of all, we call it an arcminute to make it clear that it is a unit of measurement of angles, and to distinguish it from the "common" minute, which is a measure of time. Since an arcminute is 1/60 of a degree, and we know that a circle is subdivided into 360°, this measure represents 1/21600 of a circle. However, these figures can be made intuitively understandable by giving a practical example. A common housefly is more or less 1 cm large, and its angular size would be 1 arcmin when seen from about 35 m; or, if you prefer, a 5-m long car would be 1 arcmin large from about 17 km, that is, from twice the height of Mount Everest.

In ancient times, when no telescopes, spyglasses, or other tools able to enhance the performance of human eye existed, it represented the accuracy limit of astrometric measurements. This has now been improved by several orders of magnitude, but such accuracy was largely sufficient to make important scientific discoveries, like the precession of the equinoxes.

Today, we describe this phenomenon within a physical model involving gravity and dynamics, which was first introduced by Newton. In this context, we know that a spinning body subject to a couple behaves like a top; its spin axis does not rotate following the direction of the forces, but perpendicularly to them, describing a cone.

Figure 5.1 shows the way that this applies in the case of the Earth. Our planet orbits around the Sun, but, at the same time, its spin axis, the red arrow in the figure, is tilted with respect to the orbital plane. The solar gravitational force, depicted by the green arrows, is different on the opposite sides of our planet, thus generating a couple acting upon our spinning planet. At first sight, since the force on the side closer to the Sun is stronger than the one on the opposite side, one might be tempted to think that the Earth should be pulled upright, until its spin

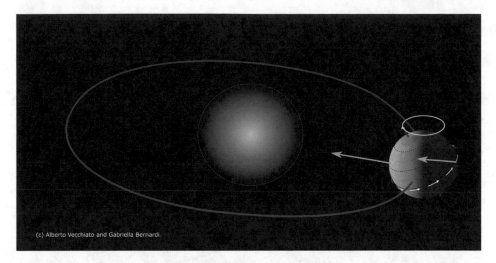

(c) Alberto Vecchiato and Gabriella Bernardi

Fig. 5.1 Representation of the precession of the Earth's spin axis caused by the influence of the solar gravitational pull, which is at the origin of the precession of the equinoxes (*Credits* Alberto Vecchiato and Gabriella Bernardi)

axis becomes perpendicular to the orbital plane. Instead, the effect of such a couple is completely different; the axis moves along the direction of the couple, which is always tangent to the yellow circle. So, from the initial position represented by the red arrow, as time goes by, the spin axis would follow the yellow circle, until it returned to the starting point after one complete turn.

As shown in Fig. 5.2, the spin axis of the Earth is fundamental when we need to determine the position of the stars, because it identifies the celestial poles, that is, those ideal points where it "pierces" the celestial sphere. This phenomenon has another interesting consequence.

We all know that the lengths of the day and the night vary periodically during the year, and that there exist two days, called equinoxes, upon which the lengths of the former and the latter are equal. The difference between the two consists in days becoming longer at the spring equinox, and shorter at the autumn one. The lengths of the day and the night coincide, because the spin axis of the Earth is perpendicular to the line connecting the sun with our planet. In this way, the poles lie exactly on the line dividing the dark and the lit sides of the globe. For this reason, the circular path followed by any point on its surface while our planet spins around its axis is split into two equal parts by such line. In any other moment of the year, the spin axis is tilted with respect to the direction of the sun, and the paths are unevenly split, so that the length of the day and that of the night are different.

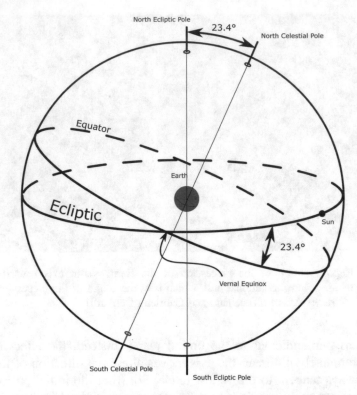

Fig. 5.2 The North and South celestial poles are those ideal points where the Earth's rotation axis intersects the celestial sphere. In one of the celestial coordinate systems, latitudes are counted north and south from the celestial equator (the plane perpendicular to the rotation axis) and longitudes are counted along the equator from the Vernal equinox. Because of the Earth's orbital motion, the Sun appears onto a background of stars that changes during the year, forming a path called Ecliptic (*Credits* Alberto Vecchiato and Gabriella Bernardi)

What does this have to do with the precession? The equinox, as mentioned, is the moment of the year when the spin axis is orthogonal to the direction of the sun, but the axis precesses in the opposite sense with respect to the orbital motion of the Earth, thus the equinox anticipates a bit at each orbit. This also explains the name adopted for such phenomenon, which comes from the Latin praecedere, meaning "to come before."

Precession was discovered in ancient times, but this discovery has little to do with farmers and seasons. Rather, it is credited to the astronomical work of Hipparchus, an important Hellenistic scientist born around 190 B.C. Although he was born in Nicaea, an ancient city located in the northwest of modern Turkey, about 100 km from Istanbul, he developed his scientific activity mainly in Rhodes, where he observed the sky for over 30 years, probably dying around 120

Fig. 5.3 It was believed that Hipparchus' stellar catalog was lost forever, until, in January 2005, Bradley E. Schaefer, an astrophysicist from Louisiana State University, discovered that part of this catalog is represented on the celestial globe of the "Farnese Atlas," a Roman statue of the 2nd century A.D., a copy of a Greek original, preserved at the National Archaeological Museum of Naples, Italy

B.C. Unfortunately, our knowledge of the work of this great scientist is very limited, because none of his works survived to the present day but one, and actually, most of what we know comes from indirect reports of his results referred to by later witnesses. One of these was Ptolemy of Alexandria, who is our main source for the attribution of the discovery to Hipparchus. According to the author of the famous Almagest—the name given by the Arabs to Ptolemy's "Mathematical Treatise"—Hipparchus prepared one of the first astrometric catalogs of all time (Fig. 5.3), which included the positions of a little more than 1000 stars, divided into six classes according to their luminosity, or apparent magnitude. His catalog had been preceded by the observations made more than a century before by two other astronomers from Alexandria, Timocharis and Aristyllus, who, probably together, realized another stellar catalog. By comparing the positions of some of the stars in his catalog with those recorded by his colleagues from the past, Hipparchus noticed that their coordinates had drifted with respect to the position of the Equinox. Today, we know that the Earth's spin axis completes one precession circle approximately every 25,800 years, which implies a shift of about 1.2 arcmin/yr. A tiny bit of movement to the naked eye (remember that 1 arcmin is the best ideal resolution of the human eye), but time comes into play here as one of astronomy's most powerful allies. However small this motion can be, it does, in fact, add up over time, and, after a century, the shift amounts to a couple of degrees, a difference that was promptly noticed by a clever astrometrist like Hipparchus!

5.3   Proper Motions

What makes stars different from other objects in the sky, like the Moon, the planets or the comets, is just one simple thing. They do not move in respect to each other. In other words, while they do move in the sense that they turn around the celestial pole once per day, such motion does not change the angular distance among them. For this reason, in antiquity, they were also called "fixed stars" so as to distinguish them from "wandering stars," which, in Greek, is πλάνητες ἀστέρες ("plànētes astéres"), from which the current "planets" originated. The most evident characteristic of these objects, in fact, is that they continuously change their positions with respect to each other and to the background stars, and it took a great deal of work to develop theories capable of predicting their motion in a reliable way.

The precession described previously showed us another way in which stars change their positions, but this is, yet again, a shift of the entire sphere, which moves rigidly with respect to the current equinox.

Nonetheless, stars do move independently of each other, and, as astronomers prefer to separate the stars' angular positions in the sky from their distance, the same holds for velocities. The rate of variation of an angular position is called "proper motion," and it is obviously measured as an angular velocity, that is, in angles per unit of time.

So, how it is possible that ancient people, who were able to measure an effect as small as the precession of the equinoxes, could have considered the stars "fixed"? The simple answer is, "because the effect of the proper motion of the stellar coordinates is much smaller than that of the precession of the equinoxes." The star with the highest known proper motion is Barnard's Star, in the constellation of Ophiuchus, whose speed in the sky amounts to 10.36 arcsec/yr. One arcsecond (1'' in symbols) is 1/60 of an arcminute, or 1/3600 of a degree, and as to the housefly in our previous example with the arcminute, it would have an angular size of 1'' at a distance of 2.1 km. So, the star with the largest proper motion moves six times slower than that which can be seen from the precession of the equinoxes. Once again, the effect on the position adds up over time, and, in 100 years, its shift would be about 17 arcmin, which is a little less than 0.3°, and Hipparchus would have been able to detect it, except... he couldn't have, because this star is not visible to the naked eye!

Stars, in fact, have different levels of brightness. This has been evident to any observer since antiquity, and, in fact, in his catalog, Hipparchus classified all of the stars into 6 classes, or magnitudes; 1 for the brightest and 6 for the faintest. When this scale was conceived, neither spyglasses nor telescopes yet existed, so 6 was the faintest object perceivable by the unaided eye, and the magnitude of Barnard's star is 9.5, way too faint for any astronomer before Galileo Galilei at the earliest.

But there are other objects with a decent proper motion that are visible with the naked eye, right? Actually, the answer is yes. The star 61 Cyg A—a star in the constellation of Cygnus, the brightest of a system of two gravitationally bound objects—has a proper motion of 5.28 arcsec/yr, which is about half of that of Barnard's star, and a magnitude of 5.21. Although it is at the faint end of the visibility range, this is the fastest star visible to the naked eye, and yet, in 100 years, one would observe a bare 9 arcmin' displacement, which is just nine times larger than the best possible accuracy attainable by the human eye!

A quite difficult task indeed, and we have to stress that the precession affects every star, so whichever our target is, we have a 100% probability of observing this effect, while, for the proper motions, we have to observe one specific star. So, poor Hipparchus would have had no chance to detect the proper motion of 61 Cyg A if its position was not recorded with sufficient accuracy in the catalog of his predecessors. Moreover, not all of the stars are visible from a given location on the Earth.

Table 5.1 shows all of the stars visible from Rhodes, where, presumably, the ancient astronomer made the vast majority of his observations, with proper motion greater than 1.8 arcsec/yr and magnitude brighter than 6, ordered by increasing magnitude, i.e., from the brightest to the faintest. This limit on the proper motion was chosen because, in 100 years, it produces a change in the position of at least 3 arcmin. Even if, in principle, this is above the accuracy of the naked eye, reaching it would have been just as impossible for the great astronomer, as other factors limited the actual accuracy of his measurements, but even admitting such a remarkable result, we can note that the list includes just 11 stars, that is, at most, 1% of his catalog. Moreover, one has to admit that the same accuracy had been reached by his colleagues from 100 years in the past in regard to at least one of these stars. It is not difficult to understand why proper motion was not discovered in those times.

So, can we conclude that the ancient Greek and Hellenistic civilizations could not have imagined anything other than fixed stars, or not? As surprising as it may be, the answer is not. Later Latin and Greek authors report that one of the motivations that pushed Hipparchus to create his great catalog, apparently, was to

Table 5.1 In order to give an idea of Hipparchus's chances of discovering stellar proper motion, we have selected stars visible from Rhodes that have proper motions greater than 1.8 arcsec/yr—which produces a shift in their position greater than 3 arcmin in one century—and with a magnitude brighter than 6, that is, the faintest objects visible to the naked eye

Name	V magnitude	Proper motion (arcsec/yr)
α Boo	–0.05	2.3
τ Cet	3.50	1.9
TYC 1472-1436-2	4.05	2.3
e Eri	4.27	3.1
ω02 Eri	4.43	4.1
σ Dra	4.68	1.8
μ Cas	5.17	3.8
61 Cyg A	5.21	5.3
HD 219134	5.57	2.1
HD 131977	5.72	2.0
HD 16160	5.83	2.3

have a reference for future studies that might reveal stellar motions, another reason for our deepest admiration for this great scientist.

His work eventually bore fruit, even if it took almost 2000 years. In 1718, Edmond Halley was able to determine the first proper motion of three bright stars, Sirius, Aldebaran and Arcturus (the latter is the common name of α Boo, listed in our table) by comparing his positions with those of the catalog of Ptolemy of Alexandria, who was largely borrowed from by Hipparchus.

5.4   Parallaxes

If, as we wrote above, the sky appears to us like a "flat" celestial sphere, how can we estimate the distances of celestial objects? Astronomers can use many techniques to achieve this difficult task, but the most direct way is the so-called "method of the parallax."

Everybody can grasp the essentials of such a method by extending an arm and looking at a finger with the left and right eyes alternatively. Its position will be projected on a different background determined by the straight line connecting the observing eye and the finger itself, which has different directions in the two cases. Moreover, the less extended the arm, the shorter the distance, and the larger the apparent shift of the finger with respect to the background will be.

This idea can be extrapolated to many different scenarios in which we can observe one object from two different points of view whose distance is called base. The angular shift of this object, as seen from the extremes of the base, is twice its parallax. It is immediately possible to understand that there exist two simple relations between the base, the parallax and the distance.

1. For a given base, the larger the distance, the smaller the parallax angle, and vice versa.
2. For a given distance, the larger the base, the greater the parallax angle, and vice versa.

This simple trick, therefore, transforms a measure of length into a measure of an angle, which is the kind of measurement an astronomer is best acquainted with.

A simple principle, however, can be extremely hard to convert into a practical technique. This is just because of the extreme smallness of the parallax angles, at least, in the case of celestial objects.

As usual, let us first make an estimation from an everyday human perspective by digging a little more into the above example of the extended arm. The average distance between two eyes is about 7.5 cm, while 50 cm can be taken as a reasonable chest-to-wrist distance. In the above jargon, the former is nothing other than the base for the parallactic displacement that we see of our finger. As shown in Fig. 5.4, trigonometry enforces a simple relation between the base b, the

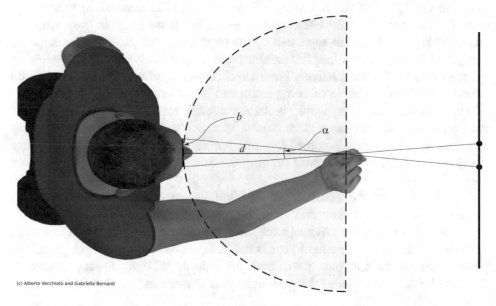

(c) Alberto Vecchiato and Gabriella Bernardi

Fig. 5.4 Schematic representation of the parallactic shift with respect to a background plane (*Credits* Alberto Vecchiato and Gabriella Bernardi)

distance d and the parallax angle $\alpha/2$, namely, $\tan(\alpha/2) = (b/2)/d$, but, even better, we do not need trigonometry at all! In our case, in fact, $(b/2)/d = 0.075$, and, for such small numbers, the tangent of an angle is almost equal to the angle itself, so we can just write $\alpha/2 = (b/2)/d$, or $\alpha = b/d$, which is the same.

This "trick," however, implies that we are not measuring angles in degrees, arcminutes, arcseconds, or any other unit introduced so far, but rather in a new unit called a radian. In other words, by alternating between the closing and opening of each eye, the finger "shifts" at an angle of about $b/d = 0.15$ radians, which means that the parallax is half of this angle, or 0.075 radians. But which is the angle in more familiar units? By definition, in the same way that a circle contains 360°, it also contains 2π radians, thus 0.15 radians correspond to about 8.6°. Quite a large angle indeed, but only because the distance is as small as 50 cm. Actually, let us take it on trust that we can still detect a parallactic shift at the resolution of the naked eye; this is a very optimistic limit, yet, with the above formulae, it is easy to compute that this corresponds to a distance of about 258 m. It is clear that, to go farther, we need to increase either the baseline or the resolution, or both; with such numbers, saying that astronomical distances are quite far is just a blatant understatement! Let us consider, for example, our closest astronomical object.

The Moon is about 384,000 km from the Earth, thus the same 7.5-cm separation between our eyes induces a parallactic shift of about 4 hundredths of thousandths arcseconds (0.00004''), namely, more than one million times smaller than the resolution of the naked eye. Such resolution is impossible to achieve, even for the most powerful telescopes on Earth, but if we imagine observing our satellite from two locations separated by the largest possible distance reachable on our planet, about 12,742 km, the parallactic shift gets back the easily detectable quantity of 1.9°. Such a relatively large effect is the reason why the parallax of the Moon was already capable of being estimated in ancient times by Hipparchus.

This baseline, however, would be barely enough for a naked-eye detection of the parallactic shift of Venus at its closest approach to the Earth, not considering that such an observation is completely impossible. Venus is our closest planetary companion, and the other planets orbiting the Sun can be up to 100 times farther away.

And what about the stars? Well, the Sun lies at about 150 million km, a distance that defines a new unit of measurement called the Astronomical Unit (AU), but, other than this obvious exception, the closest star is Proxima Centauri, 4.24 light years away. One light year is the distance covered by a light particle, a photon, travelling for one year, and the speed of light is approximately 300,000 km/s, which means that one light year is more or less 9,460 billion km.

This puts this closest star at the incredible distance of 40,110 billion km, with an approximation of a few billion, corresponding to more than 268,000 times the distance of the Sun, 800,000 times the distance of Venus, or 100 million times the distance of the Moon. As hard as it can be to accept, we have to call Proxima Centauri our "neighbor." The Milky Way is a huge disk about 1,000 light years thick and spanning across some 100,000 light years, and it is just one "town" in the Universe. Once again, except for two small satellite galaxies, the nearest neighbor of the Milky Way is the Andromeda Galaxy, at some 2.5 million light years, and there are countless other galaxies at distances that can range up to more than 10 billion light years.

Such huge figures mercilessly wipe out any possibility of using the method of the parallaxes to estimate astronomical distances, unless we make drastic improvements either on the baseline or on the accuracy of our astrometric measurement tools.

Probably, the simplest way forward would be figuring out how to exploit the orbital movement of the Earth. As 1 AU is the average distance of the Sun from our planet, this would provide an Earth-based observer with a baseline of about 2 AU, stretching from any point to its opposite on the orbital path. During the year, then, any object would describe an ellipse across the sky that is actually the projection of the orbit of the Earth onto the celestial sphere, whose dimensions are proportional to the parallax of the observed object. As can be easily understood from Fig. 5.5, the actual shape of this projection depends on the orientation of the terrestrial orbit with respect to the line of observation. For an object perfectly perpendicular to the orbital plane, the ellipse proportions are exactly those of the Earth's orbit, and it flattens while tilting down, ultimately reducing to a line when the observing direction lies exactly on the plane of the orbit.

This new baseline gives origin to another unit of measurement called parsec, which is defined as the distance at which a 1 AU baseline produces a parallactic shift of 1 arcsec. Such a unit of measurement, and its multiples, well befit astronomical needs. It implies, in fact, that the number of arcsecs contained in 1 rad is equal to the number of AU contained in 1 parsec, therefore 1 parsec is about 206,265 AU, or about 3.26 light years. The distance of the closest star, then, is a little more than 1 parsec, the size of the typical galaxies can be weighed in thousands of parsecs (Kpc) and the distances between galaxies, instead, in millions of parsecs (Mpc). Another nice characteristic of this unit of measurement is its convenient connection with the parallax, which makes it very easy to give the distance in parsecs from a parallax in arcseconds, and vice versa, because the two figures are simply the reciprocals of one another. For example, a star at 10 pc has a parallax of 1/10 of an arcsecond; an object with a parallactic displacement of 1 mas (1 thousandth of an arcsecond) can be readily placed at a distance of 1 Kpc, and so on.

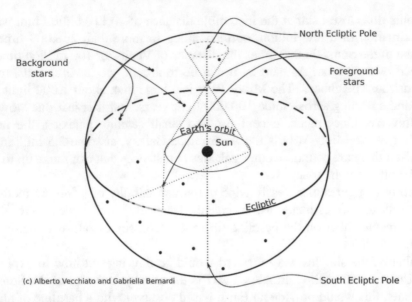

North Ecliptic Pole

Background
stars

Foreground
stars

Earth's orbit
Sun

Ecliptic

(c) Alberto Vecchiato and Gabriella Bernardi

South Ecliptic Pole

Fig. 5.5 The Earth's orbital motion induces a periodic shift of the apparent position of foreground stars in regard to the background of more distant stars. The amplitude of such motion depends on the distance of the star, and its shape depends on the orientation of the line of sight with respect to the orbit of our planet, or, in other words, on its ecliptic latitude (*Credits* Alberto Vecchiato and Gabriella Bernardi)

At the end of all of this, there is one final point that needs to be stressed about the parallaxes. When they were first introduced in this section, we vaguely mentioned the parallactic shift as the effect of a projection onto a different background caused by the changing of the point of view. But there exists no such thing as a "background," because each object is seen on the foreground of all of the objects behind it. The background concept implicitly assumes that the ellipse of parallax can be measured with respect to an abstract absolute space that, however, cannot be detected, and that does not actually exist.

What really happens is that we see the position varying with respect to other celestial objects. We know that, in principle, all of them have a finite distance, but the "ideal" ellipses of the parallax of the farther objects will be smaller than those of the closer ones, so when we measure the relative displacement of an object with respect to others that are much more distant, it will be almost the same as measuring the ideal ellipse of the parallax. This kind of astrometry, in which the position of an object is determined by measuring its placement with respect to another whose coordinates are assumed to be known, is quite appropriately called 'relative' or 'differential' astrometry. Not surprisingly, the parallax determined in this way is then called the 'relative' parallax.

The main inconvenience of this technique lies in that "almost." As shown in Fig. 5.6, in fact, the actual parallax will always be underestimated by the parallax p_0 of the reference object. This can be fixed either by acknowledging that the parallax of such a distant object is truly negligible with respect to the accuracy of measurement, or by finding a way to evaluate p_0, but, in many cases, the precision of our instruments makes these fixes very hard to achieve.

In the case of Gaia, for example, parallaxes can potentially be measured with an accuracy of up to 10 micro-arcseconds (μas). This is an incredibly small angle (10 millionths of arcseconds), corresponding to the angular size of a housefly at the distance of the Moon! A star at, say, 10,000 pc, that is, about one third of the size of the Milky Way, has a parallax of 100 μas (1/10,000 arcseconds), and measuring it with an accuracy of 10 μas means that we know this quantity with a 10% error, quite a neat measure by astronomical standards. In terms of distance, then, the relative error is more or less 1000 pc. How far should a reference object be if one wanted to reach this accuracy level with the technique of relative parallaxes?

We have to remember that the parallax of the reference object transfers directly to that of the observed one, which can have surprising consequences. One might guess that having a reference as distant as the other side of the Milky Way should be a fair request; after all, it is a distance of about 30,000 pc, three times larger

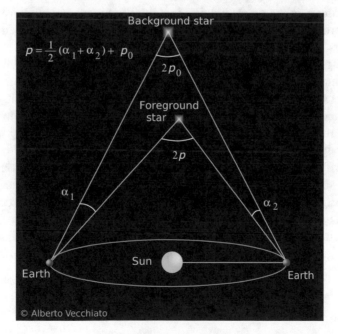

Fig. 5.6 With relative measurements, the parallax p of a foreground star is obtained by measuring the angles with respect to a much farther background object (*Credits* Alberto Vecchiato)

than that of our target object. But 30 Kpc is equivalent to 33.3 μas; with this object, this error, approximately, adds to the measurement accuracy, which means that our final accuracy would be more than 4 times worse than the one we wish to obtain. Moreover, the star will appear at less than 7,000 Kpc, instead of 10,000; more than 30% closer! Similarly, it is easy to understand now that a reference object with a parallax of 10 μas would make the accuracy twice as worse; and this p_0 corresponds to a reference object at 100,000 pc, more than three times the width of the Milky Way. If we want to leave the estimation of a 10 Kpc-distant star practically unaltered, let's say within 1%, we have to take a reference object with a parallax of 1 μas at most, that is, at 1 million pc, which is about 1.25 times our way to the Andromeda galaxy!

In conclusion, a 10% accuracy goal in distance with the methods of relative astrometry, requires, as reference, an object having a parallax of about 1% that of the observed star. The consequence is that, if the distance of the measured object doubles, that of the reference must be four times larger; if the former is 10 times bigger, the latter must increase by 100 times!

Add this to the fact that more distant objects tend to be fainter, and thus more difficult to measure, and you can easily understand why relative parallaxes, when p_0 is not known with sufficient accuracy, can be extremely hard to determine.

These difficulties can, in principle, be circumvented by adopting another approach to the parallax determination that does not need any background objects. Such an approach belongs to another branch of this discipline, called "global" or "absolute" astrometry, which we will illustrate in the next chapter.

Actually, global and absolute astrometry are two slightly different concepts, but, for the moment, it is sufficient to understand that absolute astrometry rests on the ability to measure large angles, which is a challenging task by itself, and that global astrometry requires that such measurements be done over the whole sky.

We can intuitively understand how absolute astrometry can help to determine the true parallax by looking at Fig. 5.7. The basic idea is that, if one can look contemporarily at two widely separated directions in the sky, one of which remains the same between two successive observations, then the variation of the angle between these two lines of sight can only be caused by the second source, and this variation is directly connected with the real parallax. Obviously, this is an idealized picture, useful for synthesizing a concept; the actual explanation, quite interestingly, is more general, and also has a direct connection with an Earth-based science, as we will see in a few pages.

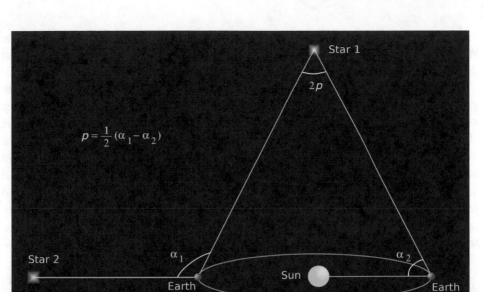

Fig. 5.7 Absolute parallaxes do not have the inconvenience of the unknown parallax of a background object, but they require the ability to measure large angles (*Credits* Alberto Vecchiato)

5.5 Perspective Acceleration

After having rummaged through the standard quantities targeted by astrometric measurements, one should note that one last parameter is missing. In the language of this discipline, positions are the equivalent, on a 2D celestial sphere, of the terrestrial longitude and latitude, while the parallax accounts for its third dimension with the equivalent of a radial distance or, if you prefer, with "height" or "depth." But the positions of the astronomical objects are not fixed in time.

This is evident for the Moon, planets, comets and in general, any object in the solar system, but, as we have seen, the same also holds true for stars. In this case, however, the largest motion, and the first to be detected in history, was the precession, which is the same for all of the celestial objects. Conversely, proper motions, that is, the movements of individual stars in the celestial sphere, could only be measured centuries later. The missing tile of our astrometric puzzle, then, would be the motion in the third dimension, which can only be the motion along the direction of observation. This, however, represents the ultimate quest for "pure" astrometric measurements, and it is so hard to detect that it was measured for the first time only in 1974 for a few nearby stars, and even Gaia will only be able to determine it for a tiny fraction of its targets!

We can intuitively understand the reason for such difficulty by imagining two different scenarios. One in which a car circles around us at a constant speed, and another in which the same car moves away from us at the same constant speed.

In the first case, two basic facts are evident. First, a constant speed on a circular path will appear to us as a constant rotation of the line of sight, namely, as a constant "angular speed" or, more correctly, a constant angular velocity, because the car, moving at a constant speed, covers equal arcs in equal times. Equal arcs in equal times are equivalent to equal angles in equal times when the distance always remains the same. Second, for identical speed, a car following a circular path at a farther distance will have an angular velocity smaller than another one closer to us. This is because both cars have to sweep 360° for a turn, but the corresponding spatial distance, that is, the perimeter of the circle, is proportional to its radius, thus the farther car will take a longer time per turn. This is the terrestrial equivalent of the proper motion, so the same considerations apply for the stars.

On the other side, the car moving away from us, following a straight path at constant speed, would initially appear to reduce its apparent (angular) dimensions very quickly, but this variation will also slow down as time goes on. As for the circular motion, the angular size of the car and its variation on time are the terrestrial equivalents of the parallax and the perspective acceleration.

Some numbers will give us a better understanding of this phenomenon. Let us take an imaginary star at 1 parsec, moving at a constant speed of 100 km/s along a straight line connecting it with the Sun, which is similar to the velocities of the typical stars in our galaxy, but with a little tuning up to make the calculations easier.

We know that, initially, its parallax is exactly 1 arcsec, but how long would it take to change the parallax from this value to 0.5 arcsec? The latter corresponds to a distance of $1/0.5 = 2$ pc, so the star has to cover a distance of 1 pc from its actual position. "By chance," 100 km/s corresponds to about 21 AU/yr, whereas we know that 1 pc is equal to 206,265 AU, thus the parallax of this star will show a variation of 0.5 arcs after some 10,000 years, approximately.

This is quite a long time, but it is also a quite large variation by current standards. So, let's ask whether, instead, we can to detect the perspective acceleration with Gaia at its best. By an act of benevolence, we are happy with just a detection, which means that it is enough to tell that the parallax has changed, but it is not necessary to measure such a change. This means that we can allow for a parallax variation that is on the order of the measurement error, namely, what would technically be called a "1-σ detection." We have already mentioned that the smallest error for Gaia is on the order of 10 μas, so the problem translates into finding how long it takes for this star to change its parallax from 1 to 1–0.00001 arcsec. Such a variation implies that the distance varies from 1 pc to 1.00001 pc, which is an increase of about 2.1 AU. This is good news, because we have seen that the star, at 100 km/s, covers 21 AU per year, so it would take just one tenth of a year for Gaia to realize that the parallax of this star had changed. The nominal mission duration is 5 years, more than enough to measure the perspective acceleration of this star. But these are just the most optimistic figures. Things change dramatically when the real case is taken into account.

There exists no star at 1 pc, but there are a certain number of them within a radius of 10 pc, so what would be the result of the same procedure for a star at this distance? At 10 pc, the parallax is 0.1 arcsec, and, as in the previous case, we want to detect a variation of 10 μas, so the new parallax is 0.09999, which corresponds to a distance of 10.001 pc. So, if, at 1 pc, the star has to cover just 0.00001 pc, or 2.1 AU, starting from a distance of 10 pc requires a shift of 0.001 pc, about 210 AU, which means 10 years at the speed of the star. Repeating the same calculations with a star at 100 pc, we need to account for 21,000 AU, and thus for 1000 years, to achieve the same parallax variation at a speed of 100 km/s.

In other words, a factor n increment in distance implies a factor n^2 increment in detectability time; a factor n of reduction of the parallax gives a factor n^2 of reduction of the perspective acceleration. Basically, this effect is as hard to detect as it is because it reduces with the square of the distance; in technical jargon, it is a second-order effect.

This also explains why an effect resulting from a constant velocity is defined as "acceleration." As in the case of the car driving away from us, the speed remains constant, but the corresponding rate of angular change decreases, and we perceive this non-constant variation in time as an acceleration, which is actually due to a perspective effect.

5.6 High-Precision Astrometry

We have said that the aim of Gaia is to map a significant part of the Milky Way, and that this requires an accuracy of measurement on the order of 10 μas. However, there are many factors that have to be taken into account to achieve such a demanding goal.

One of these factors, for example, is the atmosphere. We all owe our lives to the atmosphere, but those of astronomers, from a professional point of view, are made a little harder by this essential but imposing part of our planet. Light, in fact, is deviated from its straight path by the molecules of the atmosphere; this phenomenon is called refraction, and it is of the same nature, for example, as the one that makes a pencil plunged into a glass of water look broken at the surface of the liquid.

Refraction changes the apparent position of a star, and it obviously represents an obstacle to astrometrists. Moreover, the amount of deviation is by no means constant, but rather it varies with the inclination of the observing instrument with respect to the horizon. It amounts to zero at the zenith, but reaches tens of arcminutes close to the horizon. Moreover, the amount of refraction also depends on the temperature and pressure of the air.

The turbulence of the atmosphere disturbs astronomical observations as well, but while the effects of refraction can be, to a certain extent, modelled and corrected, the former are less predictable and of a completely different nature, because they perturb the apparent position of the observed objects in a more random way.

Another complication comes from the motion of the Earth with respect to the stars, which is at the origin of the so-called stellar aberration. Actually, in this issue, the finite speed of the light plays a fundamental role as well. This phenomenon is usually explained by analogy to a person using an umbrella in the rain.

If it is raining perpendicularly to the ground, and we do not move, then keeping the umbrella upright would protect us from the rain. However, if we start walking, the umbrella has to be inclined in the direction of motion or the rain will hit us, and the faster we move, the more we have to lean for our protection. From the point of view of an observer fixed to the ground, the rain is still falling vertically, but we are moving towards its drops, so those that we are going to intercept are not the ones over our heads, but those a little bit ahead of us, which would fall in the place that our body is going to occupy a moment later. From our point of view, instead, the rain is simply coming from a slightly bent direction, which is the result of the combination of our velocity and that of the drops.

Light is nothing other than a "photon shower" with its specific velocity, and when we observe it while moving with respect to its source—for example, a star —our velocity combines with that of the photons so that they seem to come from a different direction, and we interpret this direction as the position of the observed star. In the case of an observer orbiting the Sun, his or her velocity changes in time because of the orbital motion, thus the apparent position also changes in sync. For an Earth-based observer, the total variation amounts to about 41 arcs between two opposite points of the orbit.

Finally, gravity also plays a role when we are dealing with high-precision astrometric observations. This comes from general relativity, according to which gravity affects the motion of light in exactly the same way as it affects that of massive bodies, which has an important astronomical consequence. General relativity changes the familiar concept of a light ray moving along a perfectly straight line, and implies that, instead, it is bent when it gets close to a gravity field. The

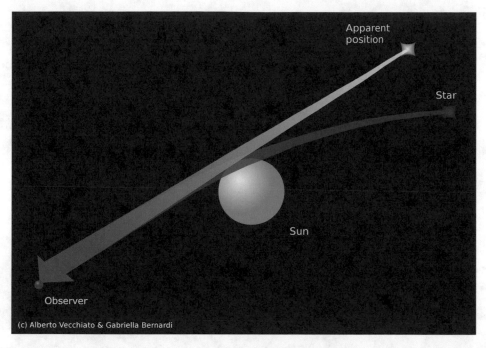

Fig. 5.8 According to general relativity, light is deflected from a straight path by massive objects. This produces a shift in the observed position of celestial objects that can be detected by high-precision astrometric measurements (*Credits* Alberto Vecchiato and Gabriella Bernardi)

closer it comes, and the stronger the field, the more pronounced the bending. This modifies the apparent direction of a star, as shown in Fig. 5.8.

The Sun generates a maximum deflection of about 1.75 arcs, which reduces to some milli-arcseconds at angular distances of tens of degrees. Other planets in the solar system have a smaller, but still consistent, effect, like Jupiter, with 16 mas at its border.

6

Global Astrometry and the Art of Celestial Map-Making

6.1 Global Astrometry from the Ground

Relative astrometry can reach a very high accuracy, but only in the usually small field visible through the telescope. If the goal is that of determining a global map of the celestial sphere, then this technique is not the best choice. It is easy to understand why by figuring out a possible approach to celestial map-making that uses relative astrometry.

One has to remember that, in this process, we do not know the coordinates of any star, and, at the same time, the choice of an origin, the point with zero longitude and latitude, is actually arbitrary. The first step, then, is choosing one star, making it the origin of the coordinate system, and determining the coordinates of all of the other stars in the field of view by measuring their positions with respect to the origin, fixing the directions of longitudes and latitudes arbitrarily in this field.

The next step is pointing our telescope in such a way that the new field of view can partially overlap the previous one. So, we can obtain the coordinates of the new stars by using those in common between the two fields as a reference. Repeat this step until the entire sky is covered and you will end up with a map of the entire sky.

This approach is simple, and correct in principle, but, as a matter of fact, unfeasible, unless we settle for accuracy at a level much lower than that achievable with modern instruments.

The main problem here comes from the fact that measurement errors, as scientists say, propagate. Let us imagine measuring the distance between two points with a 10-cm ruler having an accuracy of 1 mm. We take our measure and find a

© Springer Nature Switzerland AG 2019
G. Bernardi and A. Vecchiato, *Understanding Gaia*,
Springer Praxis Books, https://doi.org/10.1007/978-3-030-11449-7_6

separation of 5 cm, which, however, has an uncertainty of 1 mm. The meaning of this uncertainty is that the real distance is, at a given confidence level, between 4.9 and 5.1 cm. Now, we want reach another point, say, at about 12 cm from the first one and aligned with the first two. With our 10-cm ruler, however, we can only measure a 7 cm separation from the second point, and then add this to the previous one to get the total distance. But what actually is this distance? The first one, as we have seen, is between 4.9 and 5.1 cm, and the second separation will equally stretch between 6.9 and 7.1 cm, because it is measured with the same accuracy. So, the total will range between 11.8 and 12.2; its accuracy level will be twice as bad as the accuracy of the ruler, because measurement errors add up in the same way that measurements do. Obviously, the more steps you need, the worse the accuracy becomes, and vice versa; a 20-cm ruler would allow a 1-mm accuracy for a longer stretch.

This example applies to the angular measurements of a telescope, in which the amplitude of a field of view plays the role of the length of the ruler. Add this to the fact that, on the other side, generally better accuracies on the single measurements are attained by instruments with a smaller field of view, and the problem of map-making with precise relative astrometry becomes evident. Is it better to have a small field of view that can cover a limited area of the sky with great accuracy, or a larger field of view that is, however, less accurate?

Conversely, this also explains why global astrometry requires the ability to measure large angles. From the ground, this goal is usually pursued by using special telescopes, called "Schmidt telescopes," after Bernard Schmidt (1879–1935), who invented this configuration. Just to give an idea of the figures, one of the most famous instruments of this kind is the "Big-Schmidt" of Mount Palomar, USA, which has a field of view of about 36 square degrees. This amplitude is much larger than the one typical of "normal" telescopes, which can range between a few arcminutes to the order of a square degree, but it is also less than 1/1000 of the 41,253 square degrees of the entire celestial sphere.

There are techniques that help to mitigate the problem of "carrying the reference system" along on the celestial sphere, but making maps of the sky from the ground requires enduring another unavoidable difficulty. Any observer located on Earth can observe only part of the celestial sphere, and even if the share can be enlarged by exploiting the orbital movement of our planet over one year, we will always need at least two observers to cover the entire sky.

Two observers mean two instruments and two different observing conditions. The importance of this detail is that these different characteristics bring with them distinct measurement errors, which will alter the quality of the resulting map.

A map realized in this way will lack some uniformity—in astronomical parlance, it will have "regional errors"—and will look like a sort of patchwork of

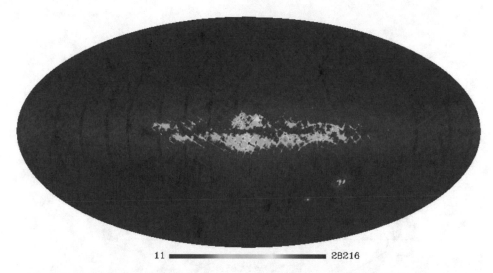

11 ■■■■■■■■■■■■■■■ 28216

Fig. 6.1 Full-sky representation of a selection of stars from the IGSL catalog used by the Gaia mission. The color scale shows the number of stars in each area. The "grid-like" feature of the map is the effect of the way the stars are identified in the different plates of the original ground-based catalogs used to produce the IGSL (*Credits* Roberto Morbidelli)

smaller charts, in which the "seams" among the patches appear as anomalies in the mapping results (Fig. 6.1).

6.2 Global Astrometry from Space and Gaia

The view of a telescope in space, unlike that of a ground-based one, has no "blind points," in the sense that, generally, no region of the celestial sphere remains permanently unobservable. This, in principle, circumvents the necessity of having at least two instruments if one wants to map the entire sky. Another advantage over Earth-based observatories is that, in space, there is no atmosphere that, as we have seen above, can degrade the accuracy of astrometric measurements because of its turbulence and the refraction effect.

Clearly, all of these difficulties disappear in space, thus an instrument able to measure large angles in the sky and placed on a satellite that can observe in any direction from above the atmosphere would, in theory, be an ideal map-making machine.

The large-angle measurements come from an idea by Pierre Lacroute (1906–1993), a French astrophysicist who proposed, in the late 1960s, the realization of a kind of "double telescope." The concept of such an instrument was to convey the observations of two fields of view onto a single focal plane. The lines of sight of

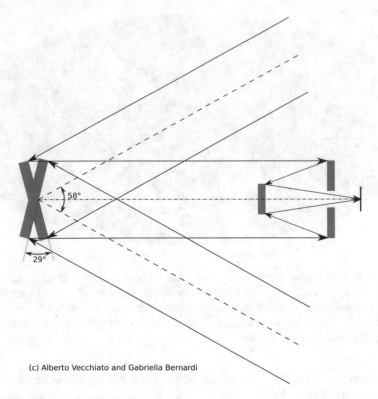

Fig. 6.2 Schematic representation of a double telescope with two fields of view separated by a base angle of 58°, as it was for the Hipparcos mission (*Credits* Alberto Vecchiato and Gabriella Bernardi)

these two fields form a large angle, called a basic angle, and the precise knowledge of this quantity makes it possible to determine the angular separation of the objects observed in the two fields of view (Fig. 6.2).

A better understanding of the way this kind of measure is capable of determining the absolute astrometric parameters across the sky requires a bit of mathematics, but the concept can be easily grasped with some intuitive reasoning.

Let us assume that our elementary measurement is the arc between two stars in the sky; for the sake of simplicity, we also want to determine nothing more than the two coordinates of the positions.

If we have two stars, then we can just take the one and only arc between them. This means that we have one measure, but 4 unknown parameters (two parameters for each of the two stars). If we add one star, then the number of unknowns increases to 6, but we can get up to three arcs out of the three stars. With four stars, instead, the number of arcs and unknowns is 6 and 8, respectively, and with five stars, we have as many stars as unknowns (Fig. 6.3). Mathematically, when

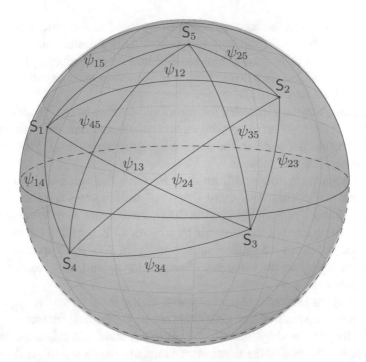

Fig. 6.3 Ideally, the sphere reconstruction process is based on a network of arcs among star pairs that, mathematically translates into a large system of equations (*Credits* Alberto Vecchiato)

the number of known quantities (the arcs, that is, the measurements) is equal to the number of unknown parameters, then we can build a system of equations that allows us to derive the former from the latter. If one has more parameters per star, it is sufficient to increase the number of stars and, consequently, the number of measures.

Obviously, this is just an ideal picture, and things do not work exactly like this. For example, this line-of-principle procedure works only if all of the arcs can be connected together exactly at their end, which means that we can make error-free measurements. But, in the real world, measures are always affected by some uncertainty, so a more correct picture would be that of a somewhat "loose" network of arcs whose endings can be adjusted a bit to get the best possible fit to the pivot points, and, in this case, the amount of "adjustment" in a certain sense defines the residual uncertainty in the determination of the unknown parameters.

The illustrative picture we have just given can also show the need to have large angles under another perspective. In principle, in fact, we could apply exactly the same procedure using short arcs, instead of the arbitrarily long ones that we have considered above, but it is intuitive to understand that the resulting network would be less "rigid," so to speak, thus admitting larger uncertainties.

Such an approach to the reconstruction of a global astrometric sphere was first applied to the astrometric mission Hipparcos. This satellite, the forerunner of Gaia, was launched in August 1989, and finished its operational phase 3.5 years later, in March 1993, while the preparation of the astrometric catalog took another 4 years and ended in 1997.

The Hipparcos catalog included the positions, parallaxes and proper motions of almost 120,000 stars, determined with an accuracy better than one milli-arcsecond (or milli-arcsecond per year, in the case of the proper motions), but this was just the main product of this mission.

More catalogs were obtained from the data from this satellite. The one named Tycho reports the positions of more than one million stars with an accuracy of about 25 mas, and Tycho-2, released in 2000, three years after the publication of the official results, extends this number by a factor of 2.5, giving the positions and proper motions of about 2.5 million stars. Finally, in 2007, a Hipparchos-2 catalog was produced regarding the same stars featured in the version of ten years earlier, but with better accuracy.

Gaia is bringing the same philosophy to a higher level of accuracy and a larger number of stars. Instead of the 120,000 stars of Hipparcos, in fact, Gaia's main catalog will be populated by more than 1 billion of stars, and it is expected that the final accuracy at the end of the mission will range between the sub-milli-arcsecond for the faintest stars and about 10 µas for the brightest ones.

The ability to measure large angles is just one of the advantages of a space-based observatory. As mentioned above, a single telescope housed in a satellite is able to observe any point of the celestial sphere, which is exactly what Hipparcos and Gaia do, although differently from what we are used to.

Typically, a telescope is used as a pointing instrument, in the sense that it is pointed at the target that has to be observed, keeping it fixed into its field of view until the end of the exposure time, like a photographic camera. This technique allows us to collect more light from the object and improve the accuracy of the measurement.

Gaia and Hipparcos, instead, are scanning instruments. Such telescopes move continuously along a predetermined path, called the "scanning law," and they never stop taking measurements. In this way, the exposure time is approximately the same for any object, which explains the difference in accuracy with the brightness of the star. On the other hand, however, a scanning telescope is better suited for the realization of a global map, because it can be easily adjusted to cover almost automatically the entire celestial sphere and allows for the observation of a huge number of objects.

The scanning law comes from the combination of three independent motions of the satellite (Fig. 6.4):

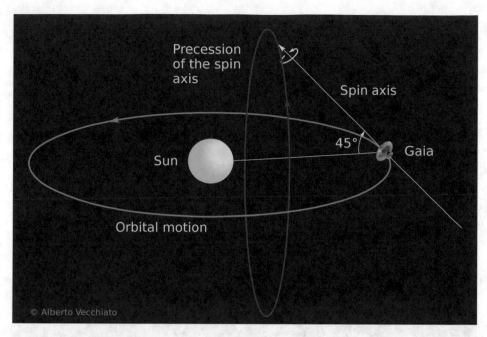

Precession of the spin axis

Spin axis

45°

Sun

Gaia

Orbital motion

© Alberto Vecchiato

Fig. 6.4 The scanning law of Gaia comes from the combination of three independent motions of rotation (spin), precession, and revolution (orbital motion) that allow the satellite to cover the entire celestial sphere about every 6 months (*Credits* Alberto Vecchiato)

1. The first one is the spinning; the two telescopes lie on a plane perpendicular to the spin axis, and thus they can scan a 0.35°-wide circular band at each turn, which, in the case of Gaia, takes approximately 6 h.
2. Actually, the scanning band is only roughly circular because of the second component, a precession of the spin axis around the line connecting Gaia with the Sun. The spin axis keeps a constant inclination of about 45°—called the 'solar-aspect angle'—with respect to the Sun-satellite direction. The precession motion of this axis transfers to the scanning circle of the spin, which therefore does not form a closed path, and thus covers different parts of the celestial sphere.
3. The above two components are not sufficient to scan the whole sky; two cones with the vertex at the satellite and axis coinciding with the Gaia-Sun direction would remain unobserved. The orbital motion of the satellite, finally, completes the scanning law and, by moving the orientation of the two cones, allows for coverage of the entire sky. The Hipparcos satellite orbited around the Earth, while Gaia is placed close to a particular point of the Sun-Earth system called the Lagrangian point L2, which is placed approximately 1.5 million km from

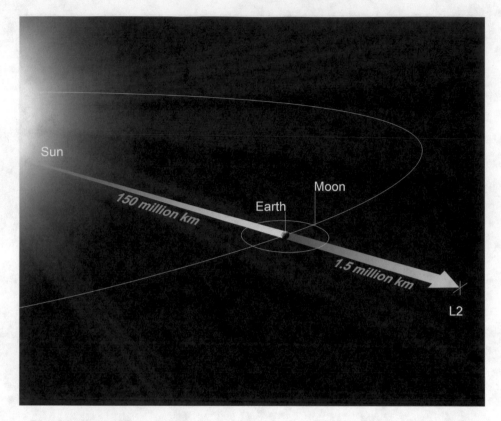

Fig. 6.5 Gaia orbits around the Lagrangian point L2 of the Sun-Earth system, located approximately 1.5 km from our planet

the Earth, on the line connecting the Sun with our planet (Fig. 6.5). For the sake of precision, Gaia is orbiting around L2, following special trajectories called "Lissajous orbits."

The combination of these three motions allows for complete coverage of the celestial sphere roughly every six months.

Part III
I as in Instrumentation and Information Technology

Photography will not be around for long.
The superiority of painting is self-evident.

Journal des savants, 1829
(French scientific journal)

7

Building the Gaia Satellite

7.1 The Spacecraft

The construction of the Gaia satellite was entrusted to EADS Astrium, now Airbus Defence and Space, and was begun in 2006 in Toulouse, France. While Astrium was the prime contractor, the development and construction of the spacecraft involved many smaller subcontractors dealing with several different parts, like the CCDs of the focal plane, which were provided by the UK's e2v Technologies, or the laser interferometers used to monitor the stability of the basic angle, which were awarded to the Dutch TNO, or the micro-propulsion system used for the attitude control, produced by the Italian branch of Thales Alenia Space. Among the 50 different companies involved in the industrial contracts, the lion's share was taken by European firms from about 15 different countries, but they also included 3 US companies. Eventually, it required more than 3.5 million working hours to study and develop this complex vehicle and its instrumentation.

Immediately after the launch, the spacecraft appears as a big 4.4-m-high cylinder with a diameter of almost 4 m, but, in its operational configuration, the satellite looks more like a big "top hat," with a base measuring more than 10 m across. Its delicate instrumentation, in fact, must operate in a thermally stable condition, with a temperature as constant as possible. This is not just for the purpose of protecting them from possible damage due to temperature variations, but also to guarantee the measurement accuracy against mechanical movements induced by such variations. The requirements, in this sense, are extremely demanding; for example, the mirrors must be kept in place to such an extent that the length of the optical path cannot change more than some tens of picometers—a length that is equivalent to the size of an atomic nucleus—for at least six hours, which corresponds to the spinning period of the satellite around its axis.

© Springer Nature Switzerland AG 2019   73
G. Bernardi and A. Vecchiato, *Understanding Gaia*,
Springer Praxis Books, https://doi.org/10.1007/978-3-030-11449-7_7

Fig. 7.1 Artistic impression of the Gaia satellite. The sunshield that protects the instrumentation from solar heating is opened at the bottom of the satellite's body, covered by its thermal tent (© ESA/ATG Medialab)

The satellite thus needs protection from the solar radiation, which is provided by a large sunshield that is initially folded around the satellite's frame. After less than an hour from liftoff, Gaia is oriented with its rear towards the Sun, and the sunshield opens up like a flower (Fig. 7.1). This fundamental component is made up of a dozen rigid frames, built with carbon-fiber reinforced plastic tubes joined by metal fittings, providing the deployable base to which the insulating material (technically called Multi-Layer Insulation, or MLI) is attached. Two parallel blankets of MLI, each constituted by a series of rectangular and triangular sections, are placed on the Sun- and shadow-sides of the structure, respectively. The rectangular ones are fixed and cover the rigid deployable frames, whereas the triangular parts are folded at the launch, and, unrolling during the opening, they fill in the space between the frames. This passive cooling system allows for a very uniform temperature to be maintained all over the several cubic meters of the satellite's volume, with variations of a few tens of millionths of a degree between one point and another of the vehicle's volume.

The same Sun that would be fatal for the Gaia instrumentation, however, is vital for its functioning, as the electric energy necessary for the satellite's operations is provided by solar panels placed on the lit rear of the spacecraft. A portion of them are fixed on the bottom of the cylinder, but additional space is allotted on the sunshield, as a dozen more are placed on each of the deployable frames. The total surface of these panels supplies an overall power of 2.3 kW at the beginning of the mission, which decreases to about 1.9 kW by the end of the satellite's operational life.

Gaia weighs more than 2 tons at launch, but this includes about 400 kg of propellant that is used for orbital and attitude maneuvers, as explained in a few lines. The rest is divided between the two parts that constitute the main body of the satellite, the Service Module and the Payload Module, which are covered by a thermal tent that gives extra shelter against the external environment.

7.2 The Service Module

This module, also known as "the platform," since the payload module is supported on it, provides all of the necessary equipment to control the functioning of the satellite, like the communication system, the central computer, the data storage, the propulsion, and other subsystems. It is comprised of the Mechanical Service Module (MSM) and the Electrical Service Module (ESM) (Fig. 7.2).

The first one is basically a structure made of Carbon-fiber reinforced plastic shaped like a hexagonal cone with a central conic hole. It houses all of the instruments of the service module and includes the propellant tanks, placed inside of the internal cone. The two propulsion systems of the satellite are also parts of the MSM: a traditional chemical system, which is more powerful and is used for the orbital maneuvers, including the insertion into the operational one and for coarse control of the satellite's attitude, and a very precise micro-propulsion system, used for the fine-constraint of the attitude around its nominal scanning law, as will be detailed in a following chapter. The thermal tent and the sunshield mentioned above are also considered parts of the MSM.

The ESM, instead, is a collection of devices, housed in the service module, that include all of the electronics needed for the satellite's operations. The Attitude and Orbit Control Subsystem (AOCS), for example, has a function that can be easily guessed by its name, but the way it was realized for Gaia is a first-run for space missions. In fact, it uses a micro-propulsion system for the fine control of the satellite's attitude instead of the more traditional reaction wheels, which had to be excluded because their movement would have disturbed the observations.

The same stability requirement that plays against a movable attitude control component influenced the way that the communication subsystem was

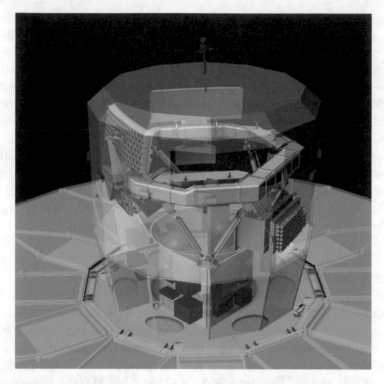

Fig. 7.2 Computer-generated view of the main body of the satellite, in which the thermal tent is made transparent so as to view the internal parts. The service module is located between the sunshield—at the bottom—and the payload module, in light blue. The components of the Electrical Service Module are colored in violet, while the orange parts represent the fuel tanks (© ESA/AOES Medialab)

implemented as well. Any satellite, in fact, is provided with one or more antennas to keep a communication link with the Earth through which scientific data, commands from the ground, and the telemetry, which, in space science, generally denotes the data that concerns the status of a space vehicle, like its position, velocity, health, and so on, can be transmitted. Gaia has three such antennas, but two of them are omnidirectional and of the "Low-Gain" type, while one is "Medium-Gain."

These prefixes are used to identify an antenna with respect to its transfer rate; generally, a Low-Gain Antenna (LGA) has a lower transfer rate than a Medium-Gain (MGA), and, obviously, both are slower than a "High-Gain" one. On the other hand, the slower version, as mentioned above, is omnidirectional, which means that it can transmit and receive signals to and from any direction, while the faster one has a specific directional link, which usually narrows as long as the downlink rate increases. You can intuitively understand the reason for this by realizing that a directional antenna is like a telescope: it concentrates the signal on a focal point, which means that it is able to detect faint sources, but it has a narrow field of view. On the other hand, an omnidirectional antenna is similar to the naked eye: it covers a large field of view, but needs a stronger signal.

Both, obviously, have their specific uses. Large amounts of data, like scientific data or complex telemetry, need a high downlink rate, whereas Low-Gain antennas can guarantee a minimum of communication even when the satellite loses its orientation with respect to the ground station. This is why Gaia's two LGAs are placed symmetrically with respect to the body of the satellite.

The MGA, however, needs to be pointed, and a steerable antenna would produce unacceptable perturbations in Gaia's measurements; just as a reaction wheel was replaced by the micro-propulsion system for the fine control of the attitude, a phased-array antenna is used instead of a steerable one. These kinds of device, as suggested by their name, are arrays of antennas in which a special device called a 'phase shifter' is placed between the transmitter and each antenna. The array of phase shifters is controlled by a computer that modifies the phase of the signal sent by the transmitter in such a way that the overall signal of the antenna, which results from the sum of that of each single antenna in the array, is a beam pointing in a specific direction. Since such a direction depends on the amount of phase shift of each antenna, which is completely computer-controlled, a directional radio wave beam can be obtained without any movable part, by just a sort of "electronic steering."

In this way, Gaia can benefit from a relatively high downlink rate for its scientific data without hampering the accuracy of its measurements. The actual rate also depends on the distance between the vehicle and the ground station and on the transmitting band; as for the latter, the microwave-range X-band is usually preferred for satellites that are not orbiting around the Earth, as in the case of Gaia, and from the roughly 1.5 million km distance of this spacecraft, the LGAs have downlink rates of some KB/s, while the MGA can reach a rate of 10 MB/s.

This transfer rate, however, is still not enough to guarantee the transmission of the huge amount of data produced by the satellite. A straight and brainless dump of every bit of information coming from the Gaia array of CCD would generate more than 35 TB of data per day, which would require a sustained transfer rate of about 3.4 GB/s every second for more than 5 years, a performance that would be hard to achieve, even on Earth! This impressive requirement can be largely reduced, for example, by selecting only a "window" of CCD pixels around each observed source; in this way, only the small fraction of "non-empty" pixels is needed. Eventually, Gaia would produce a still respectable 50 GB/day quantity of data, or about 100 TB for the whole duration of the mission, resulting in a quite comfortable transfer rate of about 4.5 MB/s. Surprising as it might seem, this is still not enough, mainly because Gaia competes for the ground stations' bandwidth with other missions. Basically, an additional reduction by a factor of 2.8 is needed to avoid losing precious information, but we cannot afford any further cuts, so how can we safeguard these valuable data?

The strategy is the same one that allows for the transmission of images, sound, or video on the internet. Broadcast companies and websites do not send the raw data, but rather their compressed version. For example, the images on a website aren't usually stored as an uncompressed bitmap, but rather as a jpeg or a png file, while broadcasting companies stream compressed mpegs rather than raw videos or WAV songs. The same is done in the case of Gaia, with the difference that we cannot use standard algorithms because they are not efficient enough with the type of data transmitted by Gaia, so new compression algorithms had to be developed. This gives rise to another complication.

Compression algorithms require a lot of calculation, and therefore adequate computational power is needed if one has to guarantee that the data can be sent within a given amount of time. You might have already experienced this kind of problem in your everyday life, for example, when trying to watch a highly compressed video on an old computer. Even if it is a file on your hard disk, and not a streaming video, modern compression algorithms are too much for older CPUs; quite often, the result is annoying frame-drops, which means that you are losing part of your video data. The same might happen to Gaia data if the satellite hardware is not powerful enough, and it is worth remembering that watching a video means that we are decoding an already compressed video, while Gaia must do the previous step, namely, that of compressing the data, which is usually the more computationally demanding of the two. Gaia can tackle this hard task only thanks to dedicated hardware with the quite intimidating name of CWICOM ASIC. The first part of the acronym stands for CCSDS Wavelet Image COMpression, and, in practice, it identifies the compression algorithm. Actually, in case you were wondering, CCSDS means Consultative Committee for Space Data System, namely, the organization formed by space agencies and associated industries that oversees the problems of communications and data system standards for spaceflight; you definitely have to get used to a lot of acronyms if you plan to work in the space sector! But the most important of the two, at least for us, is the "ASIC" part. ASIC is an "Application Specific Integrated Circuit," and, in the computer industry, it indicates, as suggested by its name, a processor designed to solve a specific problem.

A normal CPU is a sort of general-purpose computing machine: it can be programmed to solve a great deal of different calculations, so that you can use it to keep track of bank transactions, as well as operating your smartphone. Each single problem requires a different algorithm, and the software we use has the exact function of implementing a particular one into the programmable microcircuits of the CPU. An ASIC is a different fellow. It has a one-sided "mind" that allows it to solve one, and only one, problem, because the algorithm is hard-wired into its circuitry. Moreover, a CPU can be reprogrammed; the same processor that helped to compute your mortgage can be used one minute later to play some videogame. An ASIC, instead, is forever. Once you have programmed it, or, better yet, once you have built it, the program is physically embedded into its body and cannot be changed in any way.

So, why should we adopt such an inflexible device? Because flexibility comes at a cost, and stiffness can have its advantages. If it is a specific problem that matters, an ASIC can compute the same algorithm orders of magnitude faster and with less energy consumption. So, in space, where every bit of energy is precious, and with Gaia stubbornly needing to run the same program a huge number of times, a sturdy, fast and extremely efficient worker like an ASIC is extremely valuable. But, surely, as inconvenient it might be, a CPU-based solution should be possible, shouldn't it?

Actually, as a further complication, we cannot use modern CPUs in space, simply because they are too fragile. In general, every device has to be resistant to a great deal of hazards, and electronic circuits, in particular, can be easily damaged by high-energy particles outside of the protection layer of the Earth's atmosphere, to say nothing of the mere extreme temperature variations that they have to endure. It is simply impossible that a CPU designed to operate in the comfortable environment of our houses can survive in interplanetary space.

For this reason, the electronics mounted onto the spacecraft is an extremely rugged version of that which we use on Earth, or, in more elegant words, it is "space-qualified." However, it requires a lot of work and a lot of time to design and build a space-qualified electronic circuit, so the space-qualified CPUs, for example, are (in technological terms) generations behind those that we are using today. Gaia is equipped with a computer based on the ERC-32 CPU. If you have never heard of it, that's normal, firstly because this is the name of the space-qualified version of the original processor, and secondly, because its "earthling" relative was produced between 1987 and 1992. And, by the way, it is a 32-bit CPU with a clock between 14 and 40 MHz. It can more or less be compared to the old Intel 80386 or, at most, the 80486, which are roughly its contemporaries. And this was the state-of-the-art for processors in space.

7.3 The Payload Module

Fastened over the service module, and protected like the latter by the thermal tent, there is the payload module. In space science parlance, the payload is that part of the vehicle (in this case, of the Gaia spacecraft) that does the actual measurements (Fig. 7.3).

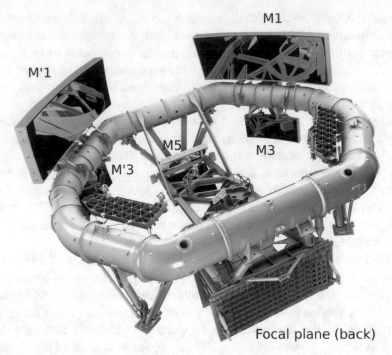

Fig. 7.3 The payload module is supported by a large SiC torus (see text) to which all of the mirrors and the focal plane instrumentations are attached. Light enters from two holes cut out of the thermal tent, and hits the primary mirrors M1 and M′1 first. A series of reflections brings it to the focal plane (© ESA—C. Carreau)

All of the measuring instruments are placed in this section of the satellite, and they are mounted on a large structure made of Silicon Carbide (SiC), constituted by a segmented loop sustained by three bipods that lie on top of the service module; this is a very stable material from a thermal point of view, which is very important for guaranteeing accurate astrometric measurements. For example, a body made of a highly conductive material like iron would be far more affected by temperature variations—which are unavoidable, because the satellite's motion changes its orientation with respect to the Sun—and this would modify the configuration of the telescope, for example, changing the distances between the mirrors, with disruptive effects on the telescope's accuracy. Moreover, SiC is also very light, which helped to reduce the weight of the satellite, and therefore the cost of the launch.

When, in the previous chapter, we described the basic concepts of global sky-mapping, we stressed that one fundamental constraint was the possibility of measuring large angles in the sky, and that, in the case of Gaia, this was realized with a sort of double telescope, able to see contemporaneously in two directions, an idea born in the late'60s of the previous century and made real for the first time with the Hipparcos satellite.

Two large rectangular holes are cut in the thermal tent. The light enters from there, and first meets two mirrors of the same shape (the primary mirrors M1 and M'1 in Fig. 7.3) placed on the opposite side, and fixed above the supporting structure. The holes, and therefore the mirrors, are separated by an angle of 106.5°, the previously mentioned "Basic Angle" that allows the satellite to point in two different directions at the same time and to fulfil the large angle requirement. This particular value was chosen because it also allows for better combinations of connections between different stars during a limited period of time; this makes possible a peculiar type of astrometric reduction that is used for monitoring the health of the instruments. From here, light rays are reflected to another pair of secondary mirrors (M2 and M'2) located on the other side of the structure, and below the tent holes; then, they bounce back to M3 and M'3, and yet another reflection brings the light to the last pair of mirrors, M4 and M'4, which are called the "combiners," because they project the light path onto a single mirror, M5. From this point on (Fig. 7.4), light rays proceed in parallel, first down to M6, and finally to the focal plane, where the detectors are housed.

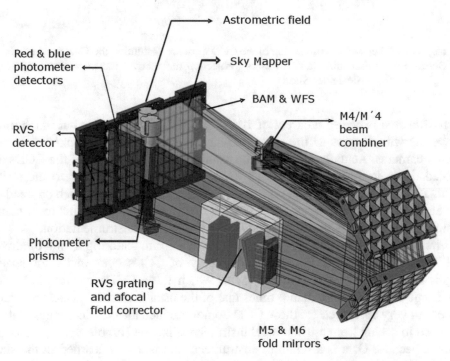

Fig. 7.4 The light coming from two different directions is combined by the M4 and M' 4 mirrors. Then, it is projected to the focal plane (on the left) by the M5 and M6 mirrors. The figure also shows the Radial Velocity Spectrometer and the prisms that intercept the light to allow spectrometric and photometric measurements (Courtesy Airbus)

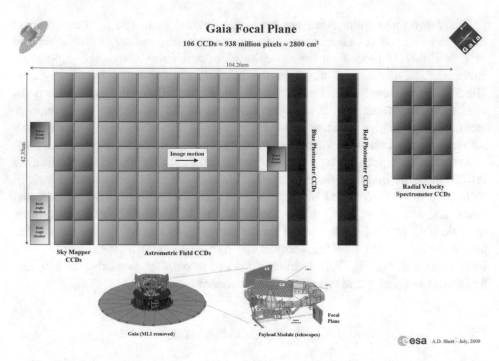

Fig. 7.5 Schematic representation of the CCD mosaic constituting the Gaia focal plane. Because of the scanning law, the stars' projection moves over the plane from left to right (© ESA—Alexander Short)

In this area is placed a carpet of 106 CCDs (Fig. 7.5), similar but much more powerful with respect to the photographic sensors of our smartphones or our digital cameras. Actually, thanks to the satellite's solar shield all of the CCDs are cooled down to 170 K, or −103.15 °C, to increase its sensibility to the stellar light. This is where the light is converted into charged particles, which are read by the electronics of the devices. But there are many peculiarities in this focal plane, which make it unique in the panorama of astronomical instrumentation.

First of all, with its dimensions of 104 cm by 42 cm, featuring a total of almost 1 billion pixels, this is the largest assembly of CCDs ever made for spatial applications. Second, the efficiency with which these CCDs convert light into electronic charge is more than 4 times that of the usual terrestrial detectors. Third, except in very special cases, these CCD do not produce the classic images that we are used to seeing from astronomical instruments like the Hubble Space telescope. This is because Gaia is a scanning instrument, that is, as we learned in the sixth chapter, it does not stare at a fixed point in the sky for an extended period, but rather it follows, without interruption, a predetermined scanning law. If ESA had placed an old-fashioned plate on the focal plane, the result would have been a series of stripes in place of the usual point-like images, similarly to what happens when we disable the following of a telescope and create an extended-exposure image. The same would happen with a CCD, if operated normally, but, in this case, this electronic device shows a great deal of flexibility, because it can be programmed to operate in a different mode, called "Time-Delay and Integration," or TDI-mode. In this way, while the satellite's motion makes stars cross the focal plane along the scanning direction, from left to right, as in Fig. 7.5, the charges produced on the CCDs are moved at the same speed in this direction, thus following the moving light spot. It takes 4.42 s to cross the entire width of a CCD, and, when, eventually, they get to the pixels at the end of it, they are discharged and the total count is recorded; in this way, we can be sure that we have not mixed up signals coming from different stellar objects. Fourth, the versatility of the CCDs is exploited in yet another way. We need to follow a typical star on its path through the focal plane to understand this point. Moving from left to right, the star first meets two columns of CCDs called the 'Sky Mapper'; a mask placed on the beam combiner M4/M′4 ensures that each column receives the light either from one line of sight or the other. In this way, it is possible to tell the objects apart from the two directions, each object is properly identified and the data of its transit are sent to the next CCDs, that is, those of the Astrometric Field (AF), which is where the next trick happens. From now on, in fact, all of the CCDs work in so-called "windowing mode," which means that the first column (AF1) confirms the identification of the object and "cuts" a window of pixels around it, and this window is propagated through the AF columns that follow. As we have already mentioned above, in fact, only that little fraction of pixels around the object is collected, so as to reduce the amount of data that have to be sent to the ground.

Last, but not least, this is not the focal plane of a single instrument, but of three instruments in one: astrometric, photometric, and spectroscopic. All of them are needed to achieve the mission's accuracy goal, but since a clearer understanding of this statement requires a somewhat longer explanation, we are postponing its details to the next section. For the moment, we concentrate on the way in which they are arranged and on their functioning by continuing to follow the stellar image on its path across the focal plane.

After the AF columns, in fact, come the BP and the RP CCDs, which are those dedicated to photometric measurements in the blue and red bands, respectively. This can happen because, before reaching the BP and the RP, the light is intercepted by two photometers, which, as shown in Fig. 7.4, are placed along the light path, between the M6 mirror and these two CCD columns. Gaia photometers are sensitive to these bands, and therefore the CCD can record photometric data, instead of astrometric.

Next, in the same way, another instrument is interposed between M6 and the focal plane; an instrument with the intimidating name of "Radial Velocity Spectrometer" (RVS). Although its actual configuration is much more complex, this device works on the same principle as a prism or a diffraction grating; it disperses the light into different directions according to its wavelength, so that one can measure the spectrum of a source, that is, the intensity of the light beam at a high number of minutely separated wavelengths. In this case, the measures are taken by a grid of 3 by 4 CCDs behind the RVS on the focal plane.

Finally, there are four more CCDs, divided into two pairs, those of the Wave Front Sensor (WFS) and the Basic Angle Monitoring (BAM) system. The former analyzes the light of the stars so as to check the optical quality of the image. Its results are able to trigger a mechanism at the level of the secondary mirrors (M2 and M′2) that can realign the telescope optics and correct any imperfection detected by the WFS. The latter, instead, is a purely metrologic instrument, and Gaia's only one, in the sense that it analyzes the light of two internal lasers, instead of that coming from the sky. These lasers send two parallel beams to the two telescopes, which are projected to the two BAM CCDs, creating an interference pattern. A variation in the orientation of the mirrors generates a variation in this pattern, which is used to measure the amount of such a change. These measurements are fundamental for the success of the mission, because as already mentioned the orientation of the mirror defines nothing other than the Basic Angle between the two telescopes' lines of sight. As we stated at the beginning of this section, the Basic Angle is fixed at an angle of 106.5°. What we haven't mentioned, but is easy to understand, is that this angle must be kept constant with a tolerance of less than a microarcsecond to avoid disruptive tampering with the astrometric measurements. Even worse, a variation of this angle can easily mimic a parallactic shift, so a Basic Angle's instability is a direct menace to the most important quantity of the Gaia catalog, the parallax. Now, in nature, there is no such thing as a fixed angle, and the BAM instrument cannot actively control the orientation of the mirrors; its function, instead, is in the "M" of the acronym, an angle that is continuously monitored and measured within the required accuracy of less than a microarcsecond. Knowing the variation of the Basic Angle at any instant allows us to take it into account when reducing the data, and to recover the original accuracy of Gaia.

In summary, by exploiting the versatility of the CCDs, and by putting appropriate instrumentation along the light path, the focal plane could be divided into separate zones in which these detectors record astrometric, photometric or spectroscopic data. ESA has conceived the Gaia payload to work as a single instrument that can carry out the job of three at the same time. But, if Gaia is an overtly astrometric mission, why saddle it with the photometric and spectroscopic measurements as well? Is such an additional burden really necessary?

7.4 Astrometry Is Too Serious to Be Left (Only) to Astrometrists

The above question, as we have already anticipated, might be answered with a short sentence. The aim of this mission is to determine the position of the objects in the sky to the microarcsecond accuracy, but this goal requires three different kinds of measurement: astrometric, photometric and spectroscopic, and their results must be conveniently combined together.

This answer, however, explains very little, and a more satisfactory explanation requires us to know something more about the kind of measurements Gaia is doing. We have dedicated almost an entire chapter to astrometry, so here, we concentrate on photometry and spectroscopy, which, basically, refer to different modes of "reading" the light that comes from the stars.

In a nutshell, photometry focuses on the quantity of overall light emitted by a given object in different bands of the electromagnetic spectrum, whereas spectroscopy analyzes that light with a more detailed resolution in regard to wavelength.

For example, the determination of the magnitude of a star is a photometric measurement, because it requires measuring its total flux of light. However, when it comes to the actual measurements, astronomers have to deal with a number of practical factors that limit the range of wavelengths in which the photons can be collected. When the measuring apparatus is located on the ground, the atmosphere quite effectively blocks frequencies in the X and Ultraviolet bands, as well as a large part of the Infrared band; moreover, the actual detectors can have different sensitivities to different parts of the electromagnetic spectrum, and filters are also used to limit artificially the range of frequencies that can enter into the telescope. For this reason, astronomers usually consider the flux only within a given wavelength range, also called the "photometric band."

Many of these bands are defined in astronomy, and the light flux from a star, and thus its magnitude, is different for each of them, but these differences do not come at random. Rather, the amount of light emitted by a star follows a very specific curve—one that was mentioned in the second chapter, called the "black body radiation curve" (Fig. 7.6)—that depends only on the temperature of the emitter. Measuring the ratio between the light fluxes at two points on this curve is enough to determine its exact shape, and thus the temperature of the star; this is equivalent to the difference between the magnitudes in two different bands, a quantity that, in astronomy, is called the "color index." This name recalls the fact that the color of an object is related to its temperature; in this way, in practice, astronomers are determining the color of a star to estimate its temperature.

In the second chapter, we also mentioned what a spectrum is, and how it can be measured by means of a prism or a diffraction grating, which separates the light of a star into its constituent frequencies, paired by a suitable recording device. In that

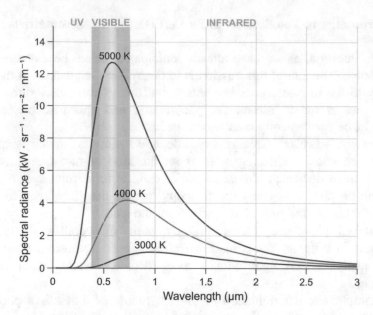

Fig. 7.6 The black body is an ideal object that is able to absorb all of the light that it receives. It re-emits in a characteristic way, represented by the black-body radiation curve, in which the peak of maximum intensity depends only on the temperature of the body. The plot shows three cases corresponding to temperatures of 5000, 4000 and 3000 degrees Kelvin, respectively

chapter, it was written that the intensity of the spectrum light at each single frequency depended on the composition of the stellar photosphere that emitted that light. With spectroscopy, astronomers determine exactly such intensities, and, in this way, they can deduce the chemical composition of the observed objects. Each chemical element, in fact, emits or absorbs light only at very precise wavelengths, so each element or molecule leaves a precise signature, in the form of a specific pattern of intensities of the spectrum.

Now, it is surely notable that stellar temperatures and compositions can be determined by photometric and spectroscopic measurements, but this does not help in understanding our initial statement, which is why Gaia's astrometric measurements have to be supported by photometry and spectroscopy to reach the desired level of accuracy.

One way in which they could help is in the estimation of the motion in the radial direction from the Sun, that is, the radial velocities of the stars. In Chap. 5, we mentioned this by saying that, in principle, it could be determined by purely astrometric measurements, thanks to an effect called "perspective acceleration." At the same time, however, we have understood that this is so minuscule that it cannot be of any help but for a tiny fraction of stars that are close and fast enough.

Spectroscopy, however, can supply this third component of motion, thanks to a phenomenon called the "Doppler effect." This expression describes an influence that motion has on sound and light waves. For the former, it is extremely easy to create an experience of it by listening to the siren of a moving ambulance. When we are at rest with respect to it, for example if we hear it from inside, it has a constant pitch; but if we are on the sidewalk and the vehicle drives past us, we would likely notice that its pitch is no longer constant. Rather, we will hear a higher note while it approaches, and a lower one when it speeds away, despite the fact that the siren, from its point of view, is emitting the same note the whole time. What happens is that the relative motion between us and the ambulance compresses the pressure waves that we perceive as sounds during the approaching phase, whereas it stretches them while the vehicle is moving away. The pitch is determined by the frequency of the wave, and, since those of the compressed ones are higher, we hear a higher note first, and the contrary happens afterwards, when lower-frequency-stretched waves reach our ears. Knowing the variation between these two notes, we can determine the velocity of the ambulance.

Since light can be interpreted as a light wave of a specific frequency, it is subjected to the same effect. In this case, however, the frequency of a light wave, when it is in the visible band, is perceived as a color, rather than a sound. Higher frequencies mean that the light emitted by the moving object is bluer, and lower ones imply a redder color. But, in the case of stars, we cannot wait until they "pass by" to see a change in their overall color to estimate their velocities. The composition of the star, however, and the ability to determine it by spectroscopy, comes to our aid in this case. As mentioned above, in fact, each element exhibits a specific spectroscopic pattern that is used to identify its presence in stellar photospheres. If a star moves with respect to the observer, such a pattern changes, because of the Doppler effect, simply by shifting rigidly towards lower or higher frequencies according to the direction of the velocity, and the degree of the shift is related by a mathematical law for such relative velocities (Fig. 7.7). One thus "just" has to identify the element responsible for a specific pattern and compare the one measured from the star to a reference one determined on Earth to give an estimation of the frequency shift, and therefore of the velocity of the star relative to the observer.

Eventually, astrometric measures came to be used to determine angular positions, parallaxes and proper motions of the observed stars; this, in the case of Gaia, involves more than a billion stars, with a final accuracy of just a few micro-arcseconds for the brightest ones and some tens of milli-arcseconds for the faintest. Moreover, spectroscopic measurements are used to determine radial velocities for objects down to about 18.5 magnitude. But why not for fainter ones?

This is because spectroscopic measurements have not been embedded in the Gaia instruments simply to provide the "sixth dimension" of the catalog; this can

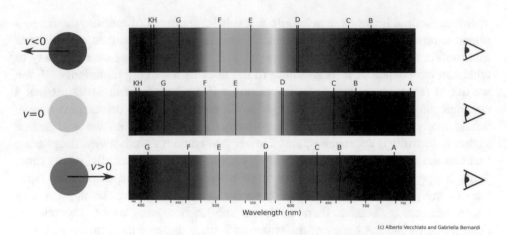

Wavelength (nm)

Fig. 7.7 The wavelength of the light coming from a moving object is shorter (blueshifted) when it is approaching, whereas it is longer (redshifted) when it recedes. This is called the "Doppler effect," and it is used to estimate the radial velocity of stars. To quantitatively determine the amount of red- or blue-shift, and thus the radial velocity, astronomers compare the spectroscopic patterns of the star's light with their equivalent measured in Earth's laboratories (see text) (*Credits* Alberto Vecchiato and Gabriella Bernardi)

be done even better from Earth, and, actually, in the case of this satellite, it is just an appreciated, though not strictly necessary, "side effect." Rather, they are needed to guarantee the final accuracy, as is the case for the photometric ones, as well as to provide other quantities of astrophysical interest, like the temperature and chemical composition we mentioned earlier.

The determination of astrometric unknowns, in fact, is done through analysis of the local positions that stars and other objects leave on the detector, like the old plates of some decades ago, or the CCDs that are used nowadays. But such positions are not pure geometrical points on a chemical or electronic canvas. They are better described as "stains," extended in size and more or less regular in shape, whose characteristics are used by scientists to extrapolate a fiducial "central location" of these spots, called a "centroid." If the shape is a regular circle, with a well-behaved intensity neatly decreasing from its center towards the border, then the centroid can be safely identified as the center of the circle, but the situation is usually not so easy.

Moreover, a detector does not react in the same way at all frequencies. This causes changes of all sorts in the shape of the spots; each kind of detector responds in its own way to an object and each telescope's configuration introduces specific perturbations, so the actual location of the centroid depends both on the combination of detector and telescope that is used for the measurement and on the "frequency composition" of the light coming from the observed object. But now we know very well that such a frequency "recipe" can be characterized by photometric and spectroscopic measurements, and, in fact, in Gaia, there are specific pipelines of the data reduction software that are devoted to determining how the instruments behave with respect to different kinds of object, and which corrections are to be applied to astrometric measurements according to that which is observed each time, a task denoted "calibration" in technical jargon.

After we have understood how Gaia collects its measurements, we need to appreciate how these data are transformed into the scientific treasure trove that astronomers from all over the world are so eager to mine.

8

On the Trail of Gaia Data

8.1 From Space to the Earth: Mission Control System

In the next chapter, we are going to see that Gaia takes its measurements from the L2 point on the Earth-Sun system, a quiet location about 1.5 million km away from the Earth. Here, the satellite collects the precious data that, first of all, have to be sent to the Earth, where the so-called "ground segment" of the mission operates.

This task is carried out by the Mission Control System (MCS), which is in charge of all of the communications with the satellite and its operations. Gaia sends a huge amount of data, scientific but also non-scientific, like, for example, those reporting the status of the instrumentation. The former can quickly fill up the storage available on board, while the non-scientific data typically require a continuous exchange of information between the satellite and the MCS. Thus, Gaia also receives a small amount of data that are needed to operate the mission; sometimes, the telescope needs to be refocused, another time, the attitude had to be changed to perform the measurements of a specific scientific test, or maybe one component has failed and a more in-depth investigation is required, but other commands can be sent for simple routine operations.

From the mere point of view of communication, an orbit around the Earth is the most convenient option, because its proximity to the ground stations allows for higher transmission rates, and, moreover, it is even possible to use a single station when the orbit is geostationary, because, with an orbital period of 24 h, the satellite will not change its position relative to the planet's surface; the easy way, however, does not befit Gaia.

Its actual location was chosen because it was necessary to keep Gaia in a very stable environment—we will return on this statement again in the next chapter—

© Springer Nature Switzerland AG 2019
G. Bernardi and A. Vecchiato, *Understanding Gaia*,
Springer Praxis Books, https://doi.org/10.1007/978-3-030-11449-7_8

whereas any terrestrial orbit is extremely ill-suited in this sense. On the other hand, this forced choice brings about some complications in the communication link.

As already said, the second Lagrangian point, or L2, of the Earth-Sun system is an ideal location about 1.5 million km beyond the orbit of our planet, around which a satellite can be orbited, keeping an approximately constant position with respect to the Earth. From our point of view, a spacecraft in L2 behaves like a star, in the sense that it rises and sets, and so it remains visible from the ground only for a limited number of hours per day.

In these cases, space agencies use a network of stations distributed over the planetary surface. ESA operates several of these ground stations, and the three most powerful amongst them, called "Deep-Space Antenna," feature 35-m large antennas that are able to keep contact with objects well beyond the Earth's orbit, like Gaia or even interplanetary probes. They are located in New Norcia, Australia

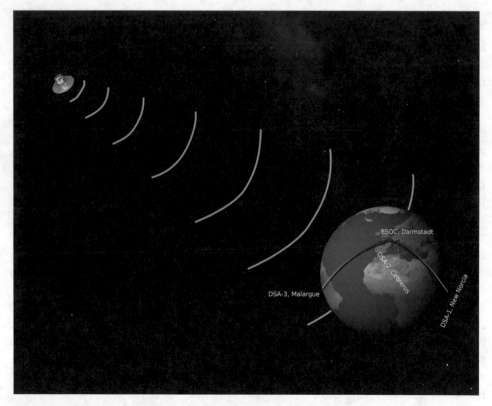

Fig. 8.1 Gaia data are first received by a series of antennas located in such a way that at least one of them can detect the signals coming from the spacecraft. From the receiving antenna, they are sent to the MCS headquarters, in Darmstadt, Germany (*Credits* G. Bernardi and A. Vecchiato)

(DSA 1), Cebreros, Spain (DSA 2) and Malargue, Argentina (DSA 3), and, thanks to their different longitudes, these powerful stations can cover the whole sky anytime.

All contact with Gaia takes place through the antennas of the Deep-Space Network, except for the delicate moments at the beginning of the mission, during the so-called Launch and Early Orbit Phase (LEOP), when additional support was provided by three more ESA stations equipped with a 15-m antenna located in Kourou, French Guinea, Maspalomas, on the isle of Gran Canaria, and Perth, Australia.

These stations are part of the MCS, and are connected to its headquarters located at the ESOC (European Space Operation Centre), a dedicated facility near Darmstadt, Germany, from which Gaia, as well as all of the other ESA missions, is controlled and operated. After their reception at one of the Deep-Space Antennas, the Gaia data are sent to ESOC (Fig. 8.1), where they are first stored in some servers dedicated to scientific data, and then sent to the next section of the ground segment, the Science Operation Centre (SOC). It is in this structure of the European Space Astronomy Centre (ESAC), near Madrid, that the processing of the satellite data starts.

8.2 Raw, Intermediate, or Well Done: The Gaia Data Processing Consortium

In this mission, a pivotal role will be played by Information Technology. When they land at ESAC, in fact, the Gaia data are simply in the form of "science telemetry," namely, they are a sort of raw data collected by the satellite instrument that need to pass through a long and careful processing phase to become usable for scientific purposes. Actually, this transformation is a task as difficult and delicate as the building of a satellite, which makes it clear why the consortium that is in charge of such processing, the Gaia DPAC (Data Processing and Analysis Consortium), is formed by hundreds of scientists from all over Europe (Fig. 8.2).

From a purely computational point of view, this processing is a formidable affair, which can be understood by looking at the computing power that comes into play. Limiting ourselves to just a couple of data, as examples, it is sufficient to mention that the Mare Nostrum machine, at the center for supercomputing in Barcelona, and Marconi, hosted at Italy's largest supercomputing center near Bologna, are involved in the Gaia data reduction process. They are two supercomputers ranking, respectively, (as of June 2018) 22nd and 18th on the list of the world's most powerful computers.

The extension and complication of this task also reflects on the practical realization of the data processing centers, which are scattered throughout different

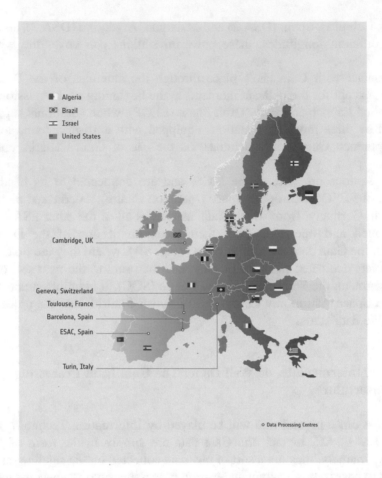

Fig. 8.2 The Gaia Data Processing and Analysis Consortium is formed by scientists coming from all over Europe and also from some non-European countries (© ESA)

European countries; there are six of them, two in Spain (the one in ESAC and another one in Barcelona) and one each in England, France, Italy, and Switzerland (Fig. 8.3).

The journey of Gaia's data through the European processing centers starts where we left them in their raw form, about 30 km west of Madrid at ESAC.

Here, an initial processing transforms the packages of raw data into an intermediate form that can be used in the subsequent stages. This has to be intended literally, in the sense that they are rearranged in a format suitable for computation, but also in a broader sense. For example, the exact observation times are computed for each observation, as well as a first estimation of their light fluxes, which are needed for determination of the magnitude. Moreover, Gaia does not know which sources it is observing, so here, the first identification happens by

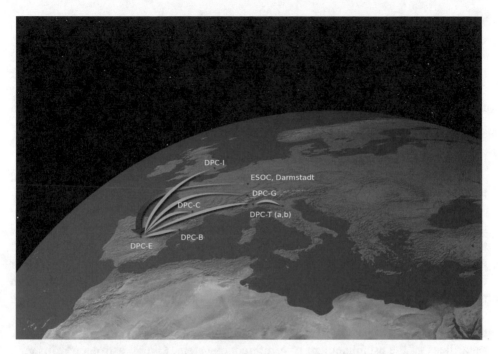

Fig. 8.3 From the MCS headquarters, data are sent to the SOC, at ESAC (see text), and, from there, they move on to other specialized Data Centers, distributed across Europe (*Credits* G. Bernardi and A. Vecchiato)

cross-matching the observations with the known coordinates, and linking them together. Finally, the first determination of the satellite's attitude, that is, of its orientation in space, is done at this stage.

Next, the data are distributed to the other processing centers across Europe for the next steps. At this point, a portion of the stars, typically some tens of million and up to about 100 million, are selected to create a first sky map, according to the procedure outlined in Chap. 6; in practice, at this stage, the reference system that is used to place all of the other stars in the sky is defined.

Data about objects with variable brightness are sent to Switzerland, where they will be classified and characterized, photometric ones find their place in England, while, in France, the pipelines for spectroscopy are run, for the identification and characterization of the objects excluded from the previous definition of the reference system, and for determination of the so-called astrophysical parameters.

In other words, at the Toulouse branch of the CNES, the French space agency, asteroids, exoplanets, double and multiple stars and all of the other objects that are not part of the previously defined reference mesh are chased. The astrophysical parameters, instead, are those quantities that are different from the plain positions and motions that have some astrophysical interest; firstly, an object is identified as a star or another kind of astrophysical object, like a galaxy, a quasar, an asteroid, or something else, then other parameters are estimated specifically for stars. They include properties like temperature, surface gravity, metallicity—which measures the amount of elements heavier than hydrogen and helium present in the stellar body—and interstellar extinction, a parameter that measures how much the light from a star is affected by the dust interposed along its line-of-sight. Like the fog on the Earth, in fact, dust makes such light dimmer and redder, and, if not properly reckoned, it can mislead the interpretation of many astronomical phenomena, as it happened in the case of Herschel's attempt of mapping the Milky Way.

In Italy, other pipelines are run, one devoted to a specific test of gravitational theories and three dedicated to a verification of similar processing already done in Spain: the estimation of the variations of the Basic Angle, the day-to-day instrumental calibrations, and the global sphere reconstruction. The reason that led to this choice is related to the so-called "absolute" character of the measurements. Gaia, in fact, is a mission that does absolute and global astrometry, the former referring to the fact that it does not just determine the positions of stars with respect to the known positions of other stars, but rather it defines a reference system. Relative astrometry, the other technique used by astrometrists, concentrates solely on measuring relative positions between two or more objects, regardless of the establishment of a reference system. Global astrometry, instead, as already mentioned in the astrometry chapter, means that such absolute measurements, and their derived reference systems, extend across the whole sky.

One can intuitively understand the difference between relative and absolute astrometry by considering the problem of measuring lengths. Relative astrometry is the astronomical counterpart of measuring the length of a specific object, whereas absolute astrometry is equivalent to determining the length of a meter, that is, of establishing the unit of measure of lengths.

It is easy to understand that, if the unit meter is wrong, all of the measurements connected with this unit will be biased, and, at the same time, one would have a hard time realizing the problem. Cross-checking with other unit meters defined in an independent way, however, would help to keep this potential problem under control. The same philosophy can be adopted in absolute astrometry.

The Hipparcos mission adopted the concept of replication to its fullest extent, establishing two consortia that processed, completely and in full independence, all of the mission data and compared the respective results before publishing the final catalog, but Gaia is too big and complex to adopt this solution, so it was decided that only some of the most delicate tasks would be replicated, specifically, the three above.

Finally, using the results that come in from the other pipelines, the intermediate data are updated and polished at the second Spanish data center, in Barcelona. When this phase is completed and the results have passed a series of verification tests to assess their scientific reliability, the astrometric catalog is released, and everything is in place to proceed to the next, and final, stage, to bring the data to their final destination.

8.3   The Endlessly Moving Data

The preparation of the computer code that is now processing the data took about ten years of work, and, indeed, it is still ongoing, because, with regard to experiments in general, and even more so with space missions, there is no dogma that enjoys more belief than this: no matter how hard you work to be prepared—you can study, you can hypothesize—when the mission starts and the real data begins streaming into your software, things will never be as you thought!

One of the most extreme cases happened with Hipparcos, Gaia's forerunner. According to the plan, the satellite would have to be placed in a circular orbit around the Earth, but, at the launch, an engine failure left the precious vehicle in an elliptical orbit completely different from that foreseen for the operations. It could have been the end of the game if the processing software hadn't been totally rewritten by people from the two Hipparcos consortia in a matter of just a few weeks!

Fortunately, the Gaia people did not have to face such havoc, although many things certainly did not go by the book. Some instruments did not behave as expected, as well as some parts of the satellite's structure that were revealed by the extreme accuracy of Gaia's measurements; even the environment in which the satellite was operating showed some unexpected features. Actually, it can be stated that Gaia's measurement helped us to discover something new about satellite manufacturing and our immediate surroundings before it even told us anything about the sky.

All of these issues had to be studied, and required appropriate modifications of the processing software, which explains why the commissioning phase, a period of the mission going from the launch to the initiation of scientific operations that is used to prepare the satellite for the operational phase, was longer than initially planned.

However uncertain the forecasts about a space mission can be, it is not a matter of dispute that the amount of data collected in Gaia's years of activity will be more than a Petabyte, corresponding to more than 1.5 million CDs, which would pile up to almost twice the height of Mount Everest.

Actually, the Petabyte scale is not actually that large, considering the quantities that we can handle today; hard disks able to store 1 TB of data, which is one thousandth of a Petabyte, are common commercial items. The real challenge of the Gaia data reduction, however, is that these data are continuously updated and the calculations made in one part of the process change the results of another part, and vice versa.…. In short, it's more like if our "Everest of data" had to be moved and shaped into the form we need; or if a director had to gauge their direction in real time according to the reactions of the public!

This rapid change is inborn to the very nature of the scanning satellite and the processing of its data. Gaia, in fact, observes the entire celestial sphere about every six months, so that one can imagine the satellite dataset as a series of huge and similar half-year-long "batches." The six-month periodicity is at the origin of the so-called "cyclic processing." Some of the reduction processes, like the reconstruction of the global astrometric sphere, require a long stretch of data to be performed, which, in principle, could follow the same periodicity of the scanning law; in practice, data are just collected over several months in large bunches, which are used to release the different versions of the Gaia catalog. At each cycle, we have new data, never processed before, but we also have new and improved versions of the data of the previous cycles. That's because the data used in these pipelines are not just those coming straight from the satellite, but also some intermediate data, produced by another kind of processing pipeline.

The telescope can defocus, a CCD can stop working, or the structure of the satellite itself can slightly move because of the continuous variation of the solar lightning. These and many other factors can degrade the performance of the instrument at any time, and cannot be foretold; the only approach is to continuously monitor and assess the behavior of the satellite in order to safeguard the quality of the measurements. This task is carried out by a series of day-to-day pipelines that constitute the so-called "daily processing." This software not only produces a sort of daily snapshot of the status of the satellite, but also uses its characterization of the instruments, technically called "calibration," to generate the intermediate data needed by the cyclic processing.

Moreover, both the calibration and the cyclic software are continuously improved and updated. It can happen, therefore, that past data are reprocessed with new versions of these pipelines. The results of a new cyclic process are thus affected not only by the presence of new data, but also by the improved quality of the intermediate data of the previous cycles and by the changes in the cyclic software itself.

8.4   Gaia Archive: The Final Destination

At the point at which we left them, some pages back, the Gaia data had been transformed from the status of raw satellite telemetry to that of scientifically meaningful sets of measures and results. All of them are stored in a Petabyte-sized database located at ESAC, the Main DataBase (MDB), with a complete copy at the Toulouse branch of the CNES and partial copies at the other data centers, the largest of which is that of the Italian data center in Turin. We might say that they are potentially usable, but not actually usable, as they have not reached their final destination, that is, the Gaia Archive.

In the main Gaia database, the data are stored and structured in a way that suits the needs of the data processing pipelines, but not necessarily that of an end user, who is, typically, an astronomer or an astrophysicist from anywhere in the world. The Gaia Archive, instead, is the vault where the data are stored with provision for their scientific use in mind.

First, relevant data are transferred from the MDB to the archive, but not yet made publicly available. Another series of validation tests that also uses data imported from other astronomical catalogs is carried out. The difference with the previous tests consists in the extent of the validation process; until now, results have been validated using only the Gaia data, at the archive, whereas, in this case, their scientific reliability is also checked against the information available from external sources.

The results of these tests provide indications as to how the actual release catalog is realized; for example, some data might have failed to reach a satisfactory level of quality, and thus have to be removed from the catalog, or scientists may have identified some peculiarities that have to be included in the catalog documentation.

A distinctive trait of the Gaia archive is that more than one release of the catalog will be published—as of this writing, two of them are already available, and at least two more will be released in the future. Moreover, mission data are usually subjected to proprietary rights, which can include a "grace period" in which they are accessible only by the scientists who have worked in the project. This is typically granted so as to repay people who have invested a considerable amount of their professional time into a work that is usually unprofitable, in terms of a scientific career. In the case of Gaia, so as to maximize the scientific outcome and the public impact, it was instead decided that data would be subject to no proprietary rights. Public access is granted as soon as they are ready, that is, as soon as they have passed all scientific validations and are properly documented.

Though it contains an impressive richness of information, the Gaia catalog is not that large, representing only a minor part of the data that will be stored in the archive. Its Petabyte-scale size, in fact, will result from the presence of the different catalog releases, but, above all, because the final release will not just include the catalog, but also the measurements of the satellite, that is, the data that was used to compute the catalog. The sheer size of the archive, however, is not so large compared to that of other projects; in the future, the huge amount of data produced by instruments like SKA or LSST in a single day will dwarf those of

Gaia, and, in 2017, the archive of the CERN, the European center for nuclear research, already contained more than 200 PB of data. Therefore, the great challenge represented by the Gaia archive does not just lie in the problem of making available some Petabytes of data to a certain number of users; the real issue here is that, in this case, these data are "live" and "interconnected."

At CERN, for example, these hundreds of Petabytes are stored in tape libraries, slow media appropriate for the goal of long-time preservation, but not for fast availability. They contain the results of archived experiments in the same way that a bank vault contains a stock of gold bars; they are incredibly valuable, but we cannot go shopping with them!

The Gaia archive is more like a collection of bank accounts connected by a credit card network; each scientist needs to have a credit card that makes it possible to use this "scientific money" to bring new ideas to reality and produce new valuables that didn't exist before. In short, the richness of the Gaia data is not just in what will be discovered by the end of the mission, with the final release of the catalog, but rather in what will be brought to light by the horde of frenzied astronomers and astrophysicists who will—and actually are—using the data.

This research will fall in the realm of Big Data archives, with lots of data transfer and processing. In such a situation, when hundreds, or thousands, of people need to use the same data, and maybe a large fraction of said data at that, the resulting activity can easily choke the network bandwidth. For the first time in a scientific environment, it will probably be better to bring the processing software close to the data, and not the other way around, a concept known as "Software as a Service" in IT. This also implies that the archive has to be conceived as, again in IT parlance, a "Service-oriented architecture," rather than a "Data-oriented" one, in the sense that, if software has to be run close to the data, the architecture must be able to provide some form of computing power as well, and this must be composed dynamically according to the needs of the moment. This is another emerging paradigm in IT technology, which uses the concept of virtualization to build "on the spot" virtual machines that are nothing other than pieces of software simulating other machines that can be assembled at will, hiding the complexity of the low-level physical machines that lie behind this layer. Moreover, the data need to be explored in an efficient way. A common way to have first contact with them is by visualizing them with plots and graphs of various quantities, but, in the case of Gaia, the data are not only too big, they are also too interconnected. There are a huge number of independent parameters to be explored, too many possible known and unknown relations between them that cannot be interpreted simply by direct observation. In other words, just asking it to "show me all of the stuff you have" would surely result in an unintelligible mess. This requires the implementation of new ideas and new tools able to filter and simplify the data before showing them off in a neat and appealing plot, another challenge that adds to the IT technologies underlying the management of the Gaia archive.

Now that we have grasped the complexity of the data processing, it is time to understand how we flew the satellite to its location and why, that is, to enter a bit into the realm of astronautics.

Part IV
A as in Astronautics and Scientific Anticipation

Lord Odysseus
Was happy as he set his sail to catch the breeze.
He sat beside the steering oar and used his skill
To steer the raft. Sleep did not fall upon his eyelids
As he watched the constellations-the Pleiades,
The late-setting Bootes, and the Great Bear,
Which men call the Wain, always turning in one place,
Keeping watch over Orion-the only star
That never takes a bath in Ocean. Calypso,
The lovely goddess, had told him to keep this star
On his left as he moved across the sea.
Homer, The Odyssey, Book V
(translated by Ian Johnstone)

9

Astronautics

9.1 Astronautics 101

Space astronomy finds itself in the particular situation of joining astronomy and astronautics, that is, one of the most ancient scientific disciplines with one of the most recent ones. We know, in fact, and we have repeatedly stressed it in this book as well, that astronomy, intended as the human inclination to look at the sky and try to understand how it works, is so ancient that we cannot date it back to a precise date, but its origins can surely be traced back to prehistoric times. Astronautics, instead, intended as that scientific and technological discipline that studies the way to travel in space, has a much more recent origin, although its basic physical principles are more ancient and can be found in the field of classical and celestial mechanics.

As a matter of fact, one can more or less arbitrarily choose the year 1883, when Konstantin Tsiolkovsky (1857–1935) wrote his first known manuscript with the initial ideas about travelling in space, or maybe May 10, 1897, when he wrote the Tsiolkovsky equation, which is at the basis of the theory of rockets, or even May 1903, when the same scientist published the first part of a work entitled 'Exploration of Outer Space by Means of Rocket Devices,' probably his most important work on the subject, in which he tackled many classical problems of astrodynamics and rocket engineering in an comprehensive and organized treatise.

Any of these dates, however, places the Russian scientist as the founding father of astronautics as a discipline with a solid scientific basis, even if, together with Tsiolkovsky, we can count among the pioneers, in order of birth year, Robert Esnault-Pelterie (1881–1957), Robert Hutchings Goddard (1882–1945), and Hermann Oberth (1894–1989).

© Springer Nature Switzerland AG 2019   105
G. Bernardi and A. Vecchiato, *Understanding Gaia*,
Springer Praxis Books, https://doi.org/10.1007/978-3-030-11449-7_9

The reason for this choice is that, although credit for the first "real" prototypical space rocket has to be attributed to Robert Goddard, in his works, this peculiar figure, both scientist and visionary, explained, for the first time, how to escape Earth's gravity with a rocket, and how the same vehicles, or at least the same principles, would have made it possible to travel in space.

One might envisage space flight as being similar to flying on Earth, and, in a certain sense, this is true, as it uses the same basic principle of physics. A propeller aircraft pushes air on its back to fly forward with a certain speed; the greater the mass of the pushed air, and the faster it is pushed, the larger the velocity of the airplane. This is a practical consequence of the principle of momentum conservation, one of the three basic laws of dynamics enunciated by Newton more than three centuries ago. The product of the mass of a body multiplied by its velocity is a quantity called momentum, and the principle states that the sum of the momenta of all of the particles of a closed system—namely, one completely isolated from external influences—must remain the same at any time. This implies that if, for example, there are two particles at rest, their initial total momentum is zero, as well as their individual one, but if, at some point, one of the two starts moving with a non-zero velocity, the other one will move with a velocity in the opposite direction. Instead of a propeller, we can also use a jet engine, in which air and fuel are mixed and burned to generate a flux of particles that is jettisoned behind the plane at high-speed.

In the atmosphere, we can vary the velocity of the surrounding air to vary our velocity in the opposite direction, but in space, there exists no air to push behind us, nor could a normal jet engine help, because it needs to suck air inside of it to burn the fuel and generate the thrust. Technically speaking, the air, or, more specifically, the oxygen in the air, is called the "oxidizer," and fuel can burn only when mixed with an oxidizer; it is the same principle that we exploit when we extinguish a fire by taking the oxygen away from it.

Tsiolkovsky thought to solve the problem of the jet engine in space simply by bringing both the fuel and the oxidizer into the vehicle. Thus, the rocket engine was born. Actual chemically propelled rockets can basically use two types of fuel, solid or liquid; for example, in liquid-fuel rockets, the oxidizer is compressed into a liquid state to increase the storage efficiency.

As strange as it may seem, however, the motion of a rocket was unexplored territory before Tsiolkovsky, again, studied it in the context of space exploration. The point is that, in physics, we use an equivalent of the law of momentum conservation to predict the dynamics of a body; it is the famous $F = ma$ law, which states that a body of mass m subject to a force F is accelerated by an acceleration $a = F/m$. The problem with this method is that it works better when the mass is constant, but we have just come to understand that a rocket works by varying its own mass, and the variation is huge for vehicles that have to escape the gravity of our planet, because almost all of the mass of a rocket launcher is propellant.

It might seem strange that, after more than two hundred years of study of Newtonian dynamics, scientists had restricted their studies almost exclusively to the problem of constant masses, but one has to remember that the most common application in astronomy was the study of the motion of bodies under the influence of gravity, which fits perfectly with the hypothesis of constant mass.

The Tsiolkovsky equation, or rocket equation, solves the problem of the dynamics of an object with variable mass, in the sense that it allows us to compute the velocity variation of a body that loses mass, as a function of the initial and final mass of the body, and of the velocity of the expelled mass.

Escaping from the Earth's gravity requires an enormous amount of energy, not only because the vehicle has to reach the impressive speed of about 11.2 km/s, but also because all of the mass of the rocket must be accelerated to this speed, thus the required energy is also increased by the need to carry around all of the propellant. This suggests a typical optimization that, yet again, was conceived for the first time by the Russian scientist, that of the multi-staged rocket. In practice, a rocket launcher is built with two or more stages, each equipped with its own engines and propellant, and, when a stage runs out of propellant, it is detached from the rocket, which can thus be pushed more efficiently by the next stage, thanks to the smaller mass.

The rocket equation is not only fundamental in order to leave Earth, but also to travel in space. Outside of our atmosphere, in fact, you always need propellant to maneuver a space vehicle, except for very particular situations. The first thing that an orbit dynamic specialist would probably try to understand when any change of route was required is its "Delta v," which is the velocity variation needed for this route correction, and that's because, through the rocket equation, this gives the required mass variation, and therefore the cost in terms of the amount of propellant. This is one of the most fundamental supplies for a spacecraft, for, when it runs out of fuel, you completely lose control over its motion. There is no friction nor breaks, so it is not like on the Earth; you need propellant both for accelerations and for decelerations, and, without it, you are simply at the mercy of gravity, which, by the way, has no mercy at all.

Actually, gravity always has to be taken into account, and it often brings about counterintuitive consequences. For example, if a satellite is in a circular orbit around the Earth and you want to move it into another circular orbit, but with a larger radius, simply pushing the spacecraft outwards is a really bad idea. What you want to do, instead, is to fire your rockets in a direction tangent to the orbit, which transforms the orbital path into an ellipse, with the farthest point at the opposite side, at the distance of the larger radius, and, once you arrive there, you have to fire again, to make the orbit circular. This is the simplest maneuver in space, which might give you an idea of the difficulties of astrodynamics, the previously mentioned branch of astronautics that studies how to move in space.

9.2 Gaia in Space

Gaia was launched on December 19, 2013, from Kourou, French Guiana, home of the main European Space Agency's spaceport since 1975 (Fig. 9.1). The location of this launch site was chosen because of its proximity to the equator, and because it has the Atlantic Ocean to the east, both very convenient features for a launching site.

In fact, an object can reach space only if it can move at a very high velocity with respect to the Earth's center, which amounts to about 11.2 km/s on the Earth's surface, namely, a little bit more than 40,000 km/h. This speed has to be obtained adding up to the velocity from the launch site with respect to the center of the planet. Such an initial velocity comes from the spin of the Earth, so, at the equator, it is largest, because, at that latitude, we are at the maximum distance from the spin axis, which implies that we have to complete a longer circular path

Fig. 9.1 The European spaceport at Kourour, French Guyana, from which Gaia was launched (© ESA—S. Corvaja)

in the same 24-h period. Thus, the equator is the place that requires the smallest amount of energy for the launch.

This is true, at least, if the launching rocket moves eastwards; contrastingly, if it moves westwards, its velocity would be opposite to the initial one, and, in this case, the equator would be the most unfavorable place to start from. This also explains why being on the east coast of the continent is also advantageous, which is simply for security reasons. Nobody wants to be standing under the trajectory of a rocket in case of a failure, and, since its trajectory is directed to the east, as explained above, having a supposedly deserted ocean underneath doesn't hurt anyone at all, literally!

Launching from the equator is also convenient in the case of satellites that need to be placed in geostationary orbits, characterized by keeping the satellite over a fixed point of the Earth's surface. These orbits are necessarily equatorial, and starting from the right latitude already allows us to avoid successive maneuvers to correct the orbital inclination, which are extremely expensive in terms of fuel usage, and therefore reduce the operating life of the satellite. Such kinds of satellite are very common, and include communication satellites, as well as those used for weather forecasts.

Gaia was launched at dawn (6:12 AM, local time) with a Soyuz/Fregat launcher. This is a Russian system composed by a Soyuz-2 "base"—a three-stage liquid-fuel rocket—paired with a Fregat rocket mounted on top of it and used as a further independent stage of the launching system. It is a thoroughly qualified rocket, already used for other space missions, for example, Mars Express.

Both components of the system use liquid propellant, but the Soyuz utilizes a fuel similar to that of jets and liquid oxygen as the oxidizer for all of its engines, including those of the four additional boosters that constitute its first stage, whereas the Fregat stage uses a hydrazine-based fuel with nitrogen-based oxidizer.

It worked as follows. In the first place, the Soyuz brought the Fregat and Gaia into a parking orbit at about 180 km over the Earth's surface (in technical jargon, a Low-Earth Orbit, or LEO); this took less than 10 min from the launch to the separation of the third stage from the "nose module" that comprises Fregat and the actual spacecraft. The two vehicles remained linked together for less than one hour, and, meanwhile, the Fregat turned its engines on twice, the first time to complete the boost into the LEO, and the second, and longer, time to place Gaia into a transfer orbit towards its final destination. At the end of this maneuver, the Fregat stage separated and moved into a graveyard orbit, while the spacecraft, now finally on its own, started its initial operations.

We are not talking about scientific operations yet, but rather all of those actions that have to be done in order to set the satellite on the right path and make it work properly. The very first step, for example, is turning on the transponder, which allows for communications between Gaia and the Earth. After about one more hour, the spacecraft is correctly oriented, with its bottom side towards the Sun and the sun shield correctly deployed, which is necessary not only to protect the payload from solar radiation and the raising of temperatures, but also to provide the energy needed for its functioning, as previously explained.

The next important phase is the placement into operational orbit. As we have already mentioned, in fact, Gaia works at the Lagrangian point L2 of the Sun-Earth system, which is located at about four times the Earth-Moon distance, or, more precisely, one and a half million kilometers from the Earth, in the opposite direction with respect to the Sun. It took about three weeks to get there, but, finally, on the 8th of January, the satellite had finished the complex maneuvers needed to enter into a stable orbit around this ideal point (Fig. 9.2). The complicated nature of these operations exists for a number of different reasons.

Firstly, the nature of the unstable equilibrium point of L2 makes the optimal path around it very different from the circular or elliptical ones that we are more used to. Rather, they are oddly-shaped quasi-periodic trajectories with

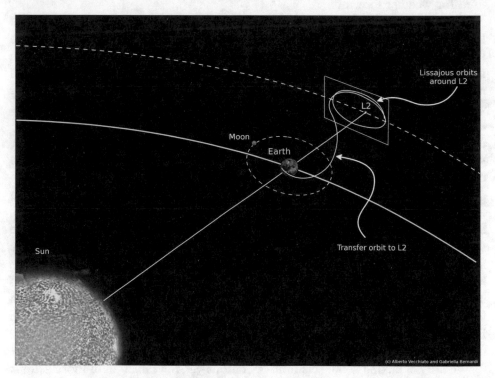

Fig. 9.2 Schematic representation of the Gaia path from the launch pad to its operational orbit, around the Lagrangian point L2 of the Sun-Earth system (*Credits* G. Bernardi and A. Vecchiato)

considerable variations in their dimension and orientation. Maybe the most striking difference with respect to a usual orbit is that Lissajous trajectories do not lie on a plane, but rather fill a three-dimensional region instead. Despite these complications, they are still unstable, in the sense that the satellite, driven only by gravity, would tend to depart from the original path, and such deviations would grow exponentially with time. Nonetheless, Lissajous trajectories require a minimum energy, and thus minimum fuel consumption, to recover these variations and to keep the spacecraft on track. This, however, is only the beginning.

The orbit has to be reached without losing the orientation of the sun shield, which is vital for the survival of the satellite. The watchword of Gaia, in fact, is "stability."

Losing the orientation might allow the Sun to hit the satellite and its instrumentation directly, making it unusable. Moreover, the antenna placed at the bottom of the spacecraft would not be able to maintain a stable communication channel with the Earth, which again means almost certain failure. Finally, the stability requirement, actually the thermal stability in this case, also plays in the opposite sense; illumination has to be neither too high nor too low, but rather as constant as possible. Now, the L2 point has the unfortunate property of falling within the penumbra of the Earth, a circle centered on the Lagrangian point with a radius of 13 thousand kilometers. Were Gaia to enter into such a zone, it would mean an untimely end for the mission, as the solar eclipse caused by the shadow of our planet would lead to a large variation of temperature and damage the instrumentation. As if that were not enough, these orbits sport another singular feature that makes the placement even riskier. Sooner or later, the satellite will enter the penumbra, so one has to select an orbit with the longest possible eclipse avoidance. It happens that the closer we start to the penumbra with a path that moves away from it, the longer it will take to get back into trouble. Quite the tightrope space walk from the very start! The sensitivity of this maneuver was explicitly stressed by the flight dynamic specialist Mathias Lauer, in an interview with the European Space Agency's blog (http://blogs.esa.int/gaia/2014/01/07/the-flight-dynamics-expertise-behind-gaias-critical-manoeuvre/): "With other missions, around Earth or Mars or even for some mission phases of Rosetta, we have essentially stable orbits and enough time to conduct the thruster burns and recover and re-plan if anything goes wrong. That's not the case with Gaia. We must carefully watch it at every step of the way."

Of course, Gaia could have been put into orbit around the Earth, which would have spared us a lot of problems and also made communication much easier, but our surroundings are far too thermally turbulent for the delicate instruments of this satellite, which needs to stay in a cold place and with very small temperature variations. For one thing, the light emission from the Earth is, quite obviously, very different from day to night, so Gaia would endure significant temperature jumps just by changing its orientation with respect to the day-to-night transition line.

In the end, ESA specialists selected a 263,000 × 707,000 × 370,000 km orbit around L2 that has an approximate period of more or less 180 days, and is guaranteed eclipse-free for more than 6 years from the time of placement. The recent extension of the mission will thus require another maneuver to change the orbit and stay away from the eclipse. The size of the orbit is also connected with another requirement; the angle that has Gaia as its origin, and the Earth and the Sun as its other vertices, cannot exceed the value of 15°, which obviously plays as a constraint for the maximum amplitude of the Lissajous trajectories.

Another interesting aspect of Gaia's astronautics is its propulsion system. Any satellite must control two kinds of motion: that of its center of mass, dealing with its trajectory, or orbit, and that of the satellite's body around its center of mass, namely, its orientation in space, also called its attitude. For Gaia, a traditional chemical engine is used for the former, while a much more precise micro-propulsion system that uses high pressure nitrogen is used for the attitude control.

In a previous chapter, it was mentioned that Gaia's scanning law is obtained by the superposition of three ideal and independent motions: the revolution around the Sun, the spin around its axis and the precession of this axis around the Sun-satellite direction. This motion can also be decomposed into an orbital and an attitude motion. In any case, this nominal scanning law is chosen to cover the entire celestial sphere approximately every six months.

Now, it is essential to the goal of the mission that the satellite is able to follow the attitude resulting from the nominal scanning law as closely as possible, but, as for the orbital motion, the satellite's body would quickly drift apart if passively left under the influence of the environment. For this reason, the attitude is actively controlled by a dozen thrusters of a special kind, the micro-propulsion system.

The closest the real attitude is to the nominal one, both in space and in time, the better, so the best thruster for this task is one that can give off small impulses very frequently; using a traditional thruster for this job would be like trying to drive a car with a jet engine. The thrusters used by Gaia, instead, are capable of giving off thrusts as tiny as 1 μN at a frequency of one per second. This is an extremely small force that can be related to a more familiar one by considering that the tiniest ant weighs about 1 mg, and therefore the weight on the Earth of one of these small ants exerts a force of about 0.01 N. Thus, Gaia thrusters are 10 thousand times more delicate! In this way, the satellite is able to keep its orientation within a few arcseconds with respect to the nominal one.

9.3   Gaia and the Astronautics of the Future

We have already seen that exoplanetary studies are starting to clear the mist around two questions that, in various forms, have intrigued and challenged mankind for a long time. Are there other worlds similar to the Earth that we can explore, or perhaps even settle? Are there other forms of life in the universe? Maybe even intelligent forms of life with which humans might establish a connection?

We have also suggested that Gaia can make a contribution to the solution of this dilemma, although we are probably still very far from a final conclusion. As it stands today, we don't know how many of the planets out there are similar to the Earth, or whether alien lifeforms or civilizations can exist, and, least of all, where, but the first seeds have started to sprout.

According to the latest researches, in fact, the formation of planets around stars is a quite common event, so that there are very likely billions of planets in the Milky Way, and, regarding the possibility of some being able to support human colonization, at least one candidate was recently found.

In 2014, the Kepler mission, a space observatory operated by NASA and devoted to the search for exoplanets with a method of transit, confirmed the discovery of a system of 5 planets around Kepler-186, a red dwarf in the constellation of Cygnus. The farthest of them, Kepler-186f, is the first extrasolar planet whose size is similar to that of the Earth (actually, it is just about 10% larger) and that, at the same time, orbits within the so-called habitable zone, namely, the region of space around a star that, in principle, allows for the existence of liquid water on the surface of a planet. We don't yet know its mass, but it might be between about 0.4 and 3.8 Earth masses, if one assumes a reasonable hypothesis on its composition. On the other hand, we know much more about its revolution period, which is about 130 days.

It would seem worth a visit, if one were ever possible. Indeed, the hard truth is that, at present, we cannot go there; there is not even the tiniest possibility of reaching this planet.

Experience has taught us that, when something becomes possible, there is a good chance that it will turn into a reality, when there is also strong motivation for its accomplishment. So, we now have an important question to answer: what is the actual character of the current difficulty? Is it just a technological problem, or a real physical impossibility?

It is quite possible that you already know the answer; nonetheless, it is worth briefly recalling it. The problem arises because of the enormous vastness of the interstellar distances, together with the limits that we have regarding speed. The law that rules over us is not that of a malevolent legislator, but rather one of nature as we currently know it. It might be that, in the future, somebody will discover other laws of physics that will allow us to triumph over this scenario, but, for the moment, this is a fact with which we must reconcile ourselves.

Kepler-186 is about 500 light years away from the Sun, which means that a light ray sent from here would take 500 years to get to that system. The speed of light is also the maximum speed permitted by the laws of physics as we know them, so this is also the smallest amount of time we would need to reach that planet.

That's simply too much if we wish to have it be explored in a reasonable period, and we have not mentioned that we are not even close to being able to reach such speeds with our technology. The fastest space probes that we have built travel at a speed less than 100 km/s, a few ten thousandths of the speed of light, and the farthest object launched from the Earth, Voyager 1, is now at about 140 AU, which is less than 20 light-hours, or 2.2 thousandths of a light year. It took more than 40 years to go that far, and we are not talking about a giant interstellar ship with a crew, but a probe the size of a small car and weighing less than a metric ton. To add insult to injury, if this probe was directed toward Kepler-186f along a direct route, it would take more or less 2.5 million years for a one-way trip, and we haven't yet talked about the problem of slowing down when arriving at the destination, as well as countless other dilemmas that one would have to solve before achieving any sensible possibility of success, regardless of time.

This very sketchy and merciless analysis seems to leave us in a hopeless situation for any interstellar travel, and yet, often disguised behind a serious scientific goal, most astronomers are probably thinking about the possibility of finding an ET somewhere around us, or at least sending off a small probe on an interstellar course. But why? And how?

First of all, it might still happen that another planet like Kepler-186f will be found orbiting around a closer star, maybe just a few tens of light years. This fact alone would not help us, given the above considerations, but there are ideas and projects that, if properly developed, could probably offer a real possibility of constructing an object able to travel this kind of distance within the time span of a human life.

For example, in 2016, two remarkable events happened that suddenly transformed an exoplanet exploration into a "mere" technical challenge. In April of that year, a board of scientists, venture capitalists, and space visionaries publically announced the Starshot project, which aims to launch a thousand centimeter-sized spacecrafts towards Alpha Centauri, about 4.4 light years from the Sun. A solar sail pushed by a set of powerful ground-based lasers would give the tiny probes a significant fraction of the speed of light, allowing at least some of them to survive the travel and reach our neighbor star in about 20 or 30 years. Moreover, in August of the same year, the European Space Observatory announced the discovery of an Earth-sized planet located in the habitable zone of Proxima Centauri, another red dwarf, which, at 4.2 light years, is even a little closer to us than Alpha Centauri.

This opportunity has little to do with Gaia, and the turbulent character of its host star probably makes this planet a worse candidate than Kepler-186f; nonetheless, the circumstance proves that we cannot exclude the possibility that, in a not-so-distant future and within the constraints imposed by our current scientific knowledge, we could be in the position of sending off a probe on a route to a real interstellar mission.

In this case, we must take for granted that such a probe will need a good map. It is, in fact, one thing to be able to travel fast and another to know where you are and how to get to where you want to go.

What about the mission design? The amount of fuel to carry with you, the provisions, the energy required to operate the instrumentation; even today, every aspect of an interplanetary mission depends on accurate planning that takes its route and duration into account, but, without an accurate map, all of this would be impossible.

We need to be able to chart a course and follow it, and, more importantly, the probe itself will need this ability, because, at interstellar distances, it will basically be on its own. There will be little possibility for communication, and even that little information would take months or years to be received, which, by the way, requires us to know where the Sun is located in order to align the antenna correctly. Therefore, we will not have a bit of a chance of piloting the probe from the Earth. The spacecraft will thus need to pilot itself; this already happens in the Solar System, but it would become a bare necessity in outer space, or our space traveler would irreparably set a course to the depth of nothingness.

10

The Science of Gaia: A Walkthrough

10.1 A Satellite Filled with Promise

In the long-term ESA program, which had to establish the planning for the space missions to be launched within the first two decades of the new millennium, Gaia was in the group of so-called "Cornerstone missions," that is, those large-budget projects deemed most important for the agency. It is easy to understand that these missions had to prevail in a highly competitive selection process against a large number of hardened contenders.

Moreover, this effort cannot be measured in mere money; as already stressed in some detail, the development and operation of this mission required the full commitment of hundreds of people, which is justified by the cornucopia of scientific results that will be made accessible by this satellite. It is thus important to understand better the scientific return of this venture, which is the goal of this chapter.

Indeed, the scientific harvest is so plentiful that it is impossible to foresee its consequences in full detail. It will probably take decades to dig into the wealth of Gaia's data, and it might even be that the most significant results are still to be imagined, hidden in the unfathomable fabric of the future, so the best we can do, for the moment, is to get a general idea by setting forth on a quick Grand Tour through the universe of the astronomers' expectations.

10.2 The Solar System

We start this virtual journey from our home, the Solar System. Here, Gaia's targets are not stars, but mainly asteroids, irregularly shaped bodies, small in size compared to the planets (Fig. 10.1). They are told apart from the other small and

© Springer Nature Switzerland AG 2019
G. Bernardi and A. Vecchiato, *Understanding Gaia*,
Springer Praxis Books, https://doi.org/10.1007/978-3-030-11449-7_10

Fig. 10.1   Ida, a relatively large asteroid of about 54 × 24 × 15 km, photographed by the Galileo probe, with its small one moon, Dactyl, in 1993

irregular objects, the comets, by means of their composition. The latter are "dirty snowballs," as they are often called, because water ice represents the lion's share of their bulk, whereas asteroids are mostly rocky or metallic objects. A small size, however, does not necessarily imply a negligible relevance, and that's exactly the case with asteroids.

First of all, the Solar System hosts a huge number of these objects, probably millions. Second, as astronomers say, asteroids are "less evolved" than the planets. The latter, in fact, because of their larger mass and gravity, have gone through a series of long and complex physical and geological transformations that altered their original structure. Scientists interested in the origin of the Solar System, therefore, are more likely to find accurate information on its primordial composition by studying asteroids, rather than their bigger relatives.

Third, they are ubiquitous. Asteroids are, in fact, grouped in "families" characterized by their location in the Solar System, by their composition, or both. The best known are those of the so-called main belt, orbiting in a wide strip between Mars and Jupiter, but there are several other of these families. Another, much larger region of the Solar System extending from the orbit of Neptune until about 50 AU is the Kuiper belt, named after the astronomer Gerard Kuiper. Like the asteroids of the main belt, the members of this family are planetesimals, that is, primordial objects formed at the origin of the Solar System that weren't able to aggregate and give birth to a new planet; on the other side, their typical compositions are very different: rocky or metallic in the main belt, "ices"—that is, elements like methane and ammonia made solid by the extremely low temperatures—in the Kuiper belt. For this reason, the latter are often dubbed simply "Kuiper Belt Objects," or KBO.

Other asteroids share the same orbit of Jupiter, more precisely, they gather in two groups, one trailing the giant planet about 60° behind its position, and another one preceding it 60° ahead. These are called "Trojan asteroids," named after the heroes of the Trojan War, and these peculiar locations are caused by the gravitational interaction of the Sun and Jupiter.

Finally, probably the most important family of asteroids, at least from our point of view, is that of the Near-Earth Asteroids, or NEAs. As their name indicates, they are the closest to our planet, and they are so important that we have even defined 4 subfamilies for their classification, named Atiras, Atens, Apollos, and Amors. The reason for their importance is clear: objects with orbits similar to that of the Earth are more likely to collide with our planet, and, in fact, the subclasses follow the same driving concept. Atiras and Amors are, respectively, asteroids whose orbits are completely inside or outside with respect to that of the Earth; both Atens and Apollos, instead, cross our celestial paths, even if they have a smaller and larger orbit with respect to that of our planet.

Celestial collisions have already happened in the past. For example, it is believed that the dinosaurs' extinction about 65 million years ago was probably caused by the fall of a 10-km asteroid; a more recent event is that which took place in Tunguska, Siberia, where, in 1908, an object of a few meters caused an explosion whose power equaled that of an atomic bomb.

The possibility of one of these objects impacting with the Earth, especially in our overcrowded condition and with societies heavily dependent on pervasive but fragile technological structures, has to be considered a global threat.

The importance of studying asteroids, therefore, is not just scientific, but very practical as well. Knowing their orbits precisely, or other physical characteristics, like size or mass, would help us not only to investigate the origin of the Solar System, but also to predict possible impacts with the Earth sufficiently far in advance to take appropriate countermeasures. The more accurate the orbits, the earlier the alert, and the easier it would be to deviate the body from its collision route.

Gaia is believed to be able to determine, by the end of the mission, the orbits of some hundreds of thousands of asteroids, as well as to estimate the mass and density of about one hundred and fifty asteroids. More importantly, this satellite can observe NEAs where no ground-based observatory can. From the Earth, in fact, it is difficult to see these objects, because, given their orbits, they can be targeted almost exclusively during daytime. Gaia, instead, has no glowing atmosphere to worry about, so it is able to follow them for a larger fraction of their celestial paths.

10.3   Exoplanets

Let us now venture out of our solar system, 600 light years away. This is the limit within which Gaia will be able to detect the presence of planets revolving around other stars, or exoplanets, thanks to the high precision of its astrometric measurements.

With its first confirmed detection dating back to 1992, the researching of extrasolar planets is a relatively recent field, and for very compelling physical reasons. A planet is much, much smaller than a star, and it does not give off light, but rather it reflects that of its star, so it is much dimmer than its companion star, like having a candle nearby a floodlight. For this reason, although not completely out of the question, direct observation is extremely hard, and has been carried out only in a few particular cases, involving planets bigger than Jupiter and very far from their mother star, using techniques that allow us to screen off the stellar light.

The detection of exoplanets, therefore, has to rely on other methods. Among them, astrometry plays a strange role; it was the first technique used for this purpose, but, at the same time, the least successful one. In fact, while, as of today, several thousand extrasolar planets have been discovered (Fig. 10.2), none of them was detected with astrometric observations. What is the reason for such an

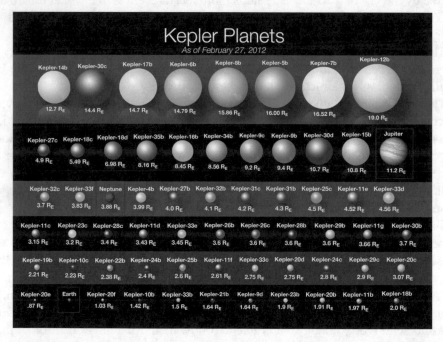

Fig. 10.2 The first exoplanets were discovered at the end of the last century. As of 9/27/2018, the number of confirmed exoplanets is 3,791 (*Source* NASA Exoplanets Archive, https://exoplanetarchive.ipac.caltech.edu/), thanks to satellites like Kepler. It is expected that Gaia will find many more (*Credits* NASA/Kepler Mission/Wendy Stenzel)

embarrassing result? And why, despite the current lack of success, are scientists still considering this possibility?

We are used to thinking that a planet orbits around a star, but this is not quite true; it might be surprising to realize that, actually, a planet pulls its mother star with exactly the same force exerted by the star on the planet. The latter, however, has a much smaller mass, and this makes the difference, because the smaller the mass of a body, the larger the velocity variation that a given force is able to transmit. This is the same principle by which, in a head-on collision between a truck and a car, the latter is thrown back much farther; in the same way, a massive star will move much less than a lightweight planet, but, in this case, the motion we are considering is rotational, rather than rectilinear. What happens exactly is that they both revolve around a point—called the barycenter, or center of mass—that lies along the line connecting the star and the planet, but such a point is much closer to the star, because the more massive body has to move more slowly (Fig. 10.3).

Now, let us imagine looking at the system from interstellar distances. We cannot see the planet, because it is too small and too faint, but its presence will be revealed by the subtle wobbling of the star orbiting around the barycenter of the system. The periodicity of the wobbling coincides with the period of the orbit,

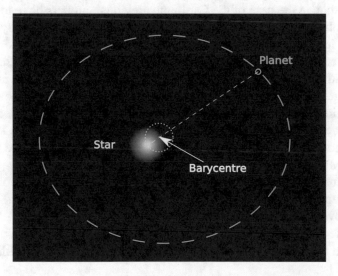

Fig. 10.3 When a star hosts a planetary system, the center of gravity (barycenter) of the system is slightly off from that of the star which, in turn, orbits around this point. This periodically changes the stellar position with respect to the background objects, thus the presence of a planet can, in principle, be detected by means of astrometric measurements. However, the radius of the star's orbit is very small, so it can be detected only with very precise astrometric measures (*Credits* G. Bernardi and A. Vecchiato)

and, together with the amplitude, this small displacement will also allow us to determine its mass, at least in principle.

The same principle has been adopted for almost two centuries to study pairs of stars that are gravitationally bound, including in the case when one of them was too dim to be seen. Actually, the first detection of these so-called astrometric binaries dates back to 1830, when the existence of the then-invisible Sirius B was deduced by the periodic variation of the proper motion of Sirius. It was then natural for astronomers to try the same with planets, and, in fact, the first attempts in this sense were made in the middle of the past century. So why, as we have already pointed out, have these efforts not been successful so far?

The problem, as usual, is in the numbers. An observer at 5 pc would observe the Sun wobbling about 1 mas around its unperturbed position, because of the gravitational pull of Jupiter. This displacement, however, requires 12 years, which is the revolution period of the giant planet. Only a few tens of stars lie within a sphere of 5 pc centered on the Sun, so we can try to estimate how far we can go with an accuracy similar to that of Gaia. The amplitude of the oscillation is inversely proportional to the distance of the star, so it would decrease to 10 μas at 500 pc, a quite optimistic limit considering that the period of the orbit is more than twice the mission's duration.

The amplitude is also proportional to the mass of the planet and to the radius of the orbit, so that, in the case of the Earth, the previous numbers have to be reduced more or less by a factor of 1500. This means less than 1 μas already at 5 pc.

In short, it is extremely difficult to discover a planet with the astrometric technique, because the actual "signature" of these bodies is very small, hardly reachable with the present instrumentation. Moreover, this method is more sensible for large planets orbiting very far from the star, but a large orbit implies a long revolution period. Jupiter's orbit has a radius of 5 AU and a period of 12 years, but Saturn is at 10 AU and has a period of 30 years. So, the large and far planets are good candidates for astrometric detection only if you are prepared to wait for a long time!

These numbers also explain why Gaia, for the first time, has a good chance of detecting exoplanets through astrometry. Current estimations tell us that it will be able to discover some thousands of Jupiter-like planets; these will help scientists to better understand the formation and evolution of planetary systems, and maybe also to obtain more data to address our questions on the existence of life outside of the Earth.

This is not the full story yet. As of October 2018, astronomers have already confirmed the discovery of almost 4000 planets, detected with different techniques (Fig. 10.2).

There exist several of them, but the two most frequently used ones are the so-called "radial velocity" and "transit" methods. The former exploits the

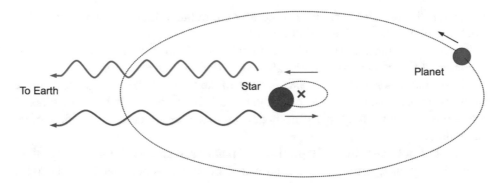

Fig. 10.4 Exoplanet Detection with the radial velocity method. The orbit of a star around the barycenter of the system provides another signature of a different kind with respect to the astrometric one. Because of the Doppler effect, the spectral lines of the stellar light are shifted towards the blue frequency when the star approaches the observer, and towards the red when it recedes. A planet whose orbit is parallel to our line of sight induces a periodic approach and recession of its star. Observing this variation in the stellar spectra allows us to detect a planetary companion and determine its mass

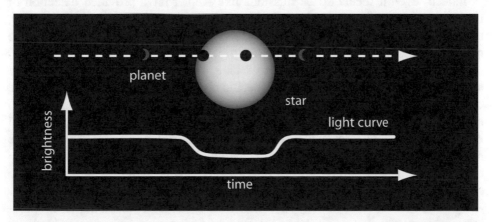

Fig. 10.5 Exoplanet Detection with the transit method. A planet, if its orbital plane is sufficiently parallel with respect to our line of sight, moves periodically between us and its star, reducing its luminosity, even if only slightly

variations in the velocity of the star along our line of sight caused by an orbiting planet (Fig. 10.4); the latter, instead, observes the variation of the stellar brightness caused by a planet passing in front of a star and behind it during its orbit (Fig. 10.5). As specified in the figures, however, both of these methods work when the orbital plane of the planet and its star is more or less parallel to our line of sight, while the astrometric technique is more sensitive to perpendicular orbits.

Moreover, the velocity variations sought by the radial velocity methods are larger when the planet is massive and close to the star, whereas we have stressed above that astrometry prefers large orbits. Finally, the transit method can infer the diameter of the planet from the brightness variation, but not its mass.

These and other features make it easy to understand that, in general, these methods are complementary to each other, so it is important to have astrometric detections, because they perform better in those exact cases in which other methods fail.

Last, but not least, the holy grail of exoplanet research is the discovery of an Earth-sized planet, orbiting in the so-called habitable zone, that is at the right distance from its star to allow for the presence of liquid water at its surface. Maybe even more ambitious is the goal of finding signatures of life in extrasolar planets. For both these goals, astronomers confide on different missions, able, for example, to analyze exoplanetary atmospheres directly.

In fact, despite its extraordinary accuracy, there's not much that Gaia can do for detecting Earth-sized planets because their astrometric perturbation on their mother stars is too small. However, this does not mean that Gaia is completely useless in this respect.

Indeed, even if recent space missions like Kepler have provided detection of many new exoplanets, and other missions are either in progress or in the works, this field of research is relatively young, and thus in tremendous need of data to make progress.

With a relatively limited amount of data, coming almost exclusively from techniques that are biased towards a specific observing configuration, the sample of Jupiter-like planets that Gaia can provide will help future studies dedicated to understanding how a planetary system can form and evolve. This is also essential in the search for terrestrial planets and extraterrestrial life; first, because these researches can provide valuable hints regarding the most promising targets, second, because it will surely help to inject more reliable numbers into the Drake equation (Fig. 2.5).

Finally, there is another way in which Gaia's data can contribute to exoplanetary science. The determination of the mass of a planet can be seriously hindered by an imprecise knowledge of the mass of the hosting star, so it is important to know this parameter, as well as being possible. It has been suggested that an estimation with a margin of error of just a few percent can be reached by combining so-called "asteroseismology," which measures the luminosity variations caused by stellar pulsations so as to investigate their internal structure, with the knowledge of the Gaia parallaxes. There are other methods to achieve this goal, but some of them are not always applicable, and, if confirmed, the previous one would give a better accuracy, so this is an example of the way in which different missions can team up to provide a new and improved solution to an old problem.

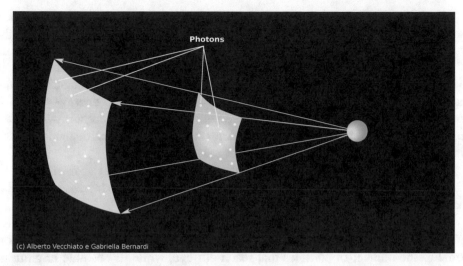

Fig. 10.6 As distance increases, the photons emitted by a star are distributed into growing areas, and therefore the source gets dimmer. (*Credits* A. Vecchiato and G. Bernardi)

10.4   Stellar Evolution

The next step brings us a little farther, up to a few ten thousand parsecs, corresponding to the size of the Milky Way. One thousand parsecs correspond to a parallactic displacement of one thousandth of an arcsecond (that is one milliarcsecond) and ten thousand parsecs to 100 μas. Since we know that the expected final accuracy of Gaia varies between 10 and 100 μas, we can reasonably say that this satellite pushes our ability to measure stellar distances all the way to the boundaries of our galaxy.

Any astronomer would cheer this news as the best opportunity to study the life of the stars. But how can the distance of a star be so important in this respect?

It all depends on the fact that, for our models of stellar evolution, the absolute magnitude of a star, or the total amount of light that it emits, is fundamental. This is a quite bold statement, but it is not difficult to understand if we remember that, in the second chapter, we discussed the stars and their life cycle. There, the H-R diagram was mentioned, stressing its fundamental importance as a tool for studying stellar evolution. In order to build this plot, we need to know the absolute luminosity of the stars, that is, the absolute magnitude.

Previously, we learned that even Hipparchus, more than 2000 years ago, had introduced the magnitudes as a way to differentiate the stars according to their brightness, so the careful reader might object that this is not much of an issue. However, the absolute magnitude that is so important for stellar evolution is not the apparent magnitude used since the time of Hipparchus.

The former is, in practice, a measure of the amount of energy emitted by a star, whereas the latter tells us how bright a star looks to our eyes and measures the amount of light that we receive. It is, again, quite intuitive to agree that the amount of energy released by a star is related to its physical status, and this is another way to get a sense of why absolute magnitude is so important to understanding stellar models. The apparent magnitude, instead, is that part of such energy—which is emitted in the form of light, in the sense of electromagnetic energy—that we receive at our observing point. The share that we can detect depends on our distance from the source.

Let's imagine an intense light, like that of the headlights of a car: it will blind us when it is close, but it becomes dimmer as long as our distance increases, until it is so faint that our eyes are not able to perceive it.

We can model light as a sort of flow of luminous particles, called photons, radiating in all directions from the source, and its apparent brightness is proportional to the number of particles that reach our eye, while its absolute brightness, or its energy, is rather proportional to the number of photons emitted from its entire surface. Therefore, for a given energy, the same number of corresponding photons will spread along a larger and larger spherical surface as they move away from the source. The number of these photons collected by a detector with a given area, therefore, will decrease with the distance from the source, and thus its luminosity will decrease as well.

To be more precise, the fraction of photons picked up by a detector (such as our eyes, a plate or a CCD) is equal to the ratio between the area of the detector and that of a sphere with a radius equal to the distance of the source. The area of a sphere increases with the square of its radius, so the apparent brightness of a star decreases with the square of its distance from the observer (Fig. 10.6).

We can consider this reasoning the other way around. If the apparent luminosity, or apparent magnitude, of a star is known, and if its distance can be determined, then it is possible to calculate the absolute magnitude.

As we know, the most precise way to estimate astronomical distances is measuring the parallaxes, and, before Gaia, we knew this quantity only for a small fraction of the stars in the Milky Way, up to a few hundred parsecs. That is why Gaia represents a much-expected revolution in this field.

10.5   Evolution of the Milky Way

We can go even further, to the borders of the Galaxy, into that approximately spherical area called the "halo" and populated by thousands of compact clusters of stars, called "globular clusters" (Fig. 10.7). Gaia will be able to determine their proper motions, and those of the spiral arms of the Milky Way, with great precision, which will provide a better understanding of the history of our galaxy.

Fig. 10.7 Globular clusters are made up of a large number of old stars arranged in spherical structures, as we see in this image of the M80 cluster, in the constellation of Scorpio

Fig. 10.8 The Mouse Galaxies (NGC4676) in the constellation of Coma Berenices show what it happens when two galaxies collide. Because of the different gravity pull among the various parts, the galaxies stretch out, forming long strips of stars. Eventually, if the galaxies merge, they will form sorts of "stellar rivers" within the resulting galaxy

Reconstructing the evolution of the Milky Way from these data is a very complex and intriguing task; the basic principles, however, are very simple. Imagine that, in a remote past, our galaxy collided with another one; such an event might seem highly improbable, but our telescopes have already photographed collisions between galaxies. Obviously, these images are not like a video of a car accident—we don't see two galaxies slowly approaching each other until they merge in a messy mix-up of stars and gases; the time scales of such astronomical events play out over millions of years, so the pictures we have are just snapshots from which the interaction of the two island-universes is apparent. What can happen after the collision, however, is difficult to figure out exactly. One might think that, after millions or billions of years, a galactic smash-up could have resulted in a confused "soup," in which there is no way to understand which star belongs to which of the two galaxies, but, as incredible as it may seem, we can!

As anticipated, the signature of the merging event is impressed in the stellar proper motions, thanks to an effect with which we are very familiar: the tides. The Earth and the Moon, although their solid bodies might seem unaffected by this phenomenon, are deformed by their mutual gravitational attraction. This is because the gravitational attraction between two bodies is inversely proportional to the distance squared, which implies that the force between the closer parts is larger than that between the more distant ones. The net effect is that bodies are "stretched" because of this imbalance in the gravitational interaction between its "front" and "rear" parts.

Internal forces that hold the different parts of the body oppose the stretching, so that the effect is minimal in solids, in which the internal forces are stronger, but liquids are much more deformable, and this results in the tides. Along the Earth-Moon junction—and, actually, also along the Earth-Sun line—the seas and the oceans stretch, causing high tides, this flow "steals" water from perpendicular directions, where the level lowers, causing low tides.

The gases and stars that make up galaxies are even less bound than the molecules of a liquid, so here the tidal stretching is even stronger, and the galaxies, as they approach each other, start to lengthen, and their structure deforms and rips. Eventually, their stars tend to arrange themselves in long filaments (Fig. 10.8).

Depending on the specific dynamics of the collision, then, galaxies can melt or separate after exchanging dust, gases and stars. If one object is much smaller, then it is likely to get disrupted by the tidal forces of the larger one, which might, contrastingly, be little affected.

Anyway, the basic idea is that these stretched structures of stars, like astronomical strings of pearls, are aligned because they are falling in the same direction and at similar speeds. This common motion can retain its identity for a long time after the collision.

Fig. 10.9 Henrietta Leavitt discovered the existence of a class of stars whose brightness varies regularly between a minimum and a maximum. The period of this variation is related to the absolute brightness of these stars, called Cepheids, after the name of the Delta star in the constellation of Cepheus. They are very bright, and therefore can also be observed in other galaxies

This is why stars with distinctly similar proper motions can be the signature of ancient galactic collisions that are still conserved in the fabric of the Milky Way. The same techniques can help us to understand how the spiral arms of our galaxy were formed, or why the stars are arranged in different ways according to their age.

10.6 Beyond the Milky Way

All of the topics discussed so far have one thing in common: our galaxy. But Gaia will also be very useful to scientists studying other galaxies and their components, and this, basically, for the same old reason: because it will be able to measure distances very well. Let's take stock and recall what we have learned.

We know that Gaia allows us to determine the distances of objects within our galaxy to a sufficient degree. Our nearest neighbor, in a galactic sense, and after our satellite galaxies, like the small and large Magellanic Clouds, is Andromeda, two and a half million light years away from us. Such a distance corresponds to a parallactic shift of 1.3 µas, too much, even for Gaia.

Moreover, the universe is full of galaxies up to billions of parsecs in distance but... we have just allowed that we will not be able to measure astronomical distances greater than some thousands of parsecs, not even with Gaia, so how is it that we can already estimate such enormous distances? That's because astronomers have come up with many other ways to calculate distances. We have already mentioned, very briefly, one of these methods in the second chapter, when we wrote about the Cepheids.

This name identifies a specific type of young star, and derives from the very first star of this type discovered in our galaxy, δ Cephei. They are variable stars, that is, stars whose luminosity changes over time, and they are also periodic, which means that these changes are regular: the brightness of the star goes from a minimum to a maximum, and again to the minimum, and so on, within a consistent period of time. At the beginning of the 20th century, the American astronomer Henrietta Leavitt (Fig. 10.9) realized that the period of variation was related to the absolute magnitude of the star.

Not long before, we mentioned that, from the apparent magnitude and the distance, we can calculate the absolute magnitude of a star; with the Cepheids, the opposite happens: we can measure the (variable) apparent magnitude of the star, and its period of variation, deducing its absolute magnitude from these data, and then, reversing the previous logic, calculate the distance!

That's great, but if we can already measure the distance with the parallax, why should we use this trick? Actually, we already know the answer. Cepheids were

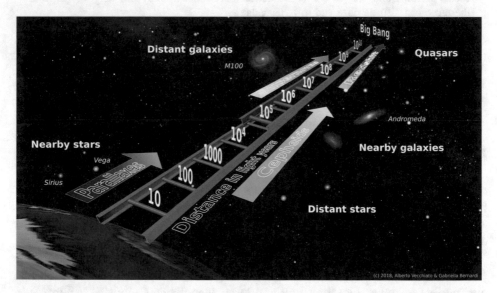

Fig. 10.10 The distances of celestial objects are determined with different techniques, each useful within a specific interval. At the base of everything, and for the nearest objects, we use the parallax (*Credits* A. Vecchiato and G. Bernardi)

first used to determine the distance of globular clusters at the beginning of the 20th century, when it was simply inconceivable to measure the parallaxes of such a faraway object.

The approach is the same now; we can use them to estimate the distance of galaxies that are too remote for the parallax method, as we just need to spot a Cepheid in that galaxy. This can work because these stars are generally very bright, and so they remain visible for a very long stretch.

We are left with just one problem now: how can we determine the relationship between the period and the absolute magnitude of the Cepheids? Since we only know the apparent magnitude variation, what we need first is to evaluate the distance of some of these stars; the period-absolute magnitude relation can then be worked out from this set of values. In technical parlance, the period-luminosity relation must be calibrated before you can apply it.

Clearly, the more precise the calibrating distances, the better the calibration, so the only reliable method is to find Cepheids that are close enough for us to use the parallax method. When you get to the real problem, things are never as simple as in a matter-of-principle exposition. Many factors are often neglected for the sake of simplicity, or maybe just because they are unknown. One of them, in this case, is that the relation that we are seeking does not involve only period and luminosity, but also the so-called color of the stars (as already mentioned, a star emits light with different intensities at different frequencies; our detectors are always sensitive to a specific range of frequencies, so the stellar apparent magnitude always refers to a given band, and the color is defined as the difference between the magnitude in two of these bands). Another one is that measurements are always affected by some kind of error, which eventually spoils the quality of the calibration, and, in order to minimize such errors, one typically wants to have a larger sample of measurements, which means more stars, possibly with a large spread in terms of distance and colors.

This is exactly why Gaia represents an unprecedented opportunity for a qualitative leap in the calibration of the Cepheids. We cannot find many of them in our neighborhoods, but, by expanding the range up to the borders of the Milky Way, we will be able to improve the number and the quality of our calibration sample by orders of magnitude.

Cepheids are just one example of what astronomers call "standard candles," that is, bright bodies whose absolute magnitude can be estimated from some kind of observable quantity (like the variability period in this case) without having to know the distance beforehand.

Other examples are another type of variable star, called RR-Lyrae, and the supernovae, whose luminosity variation during their explosion is also related to their absolute magnitude.

Having different types of standard candle is crucial for the problem of astronomical distances, because, as happens for the Cepheids with respect to parallaxes, each type has its own specific range of application. Cepheids and RR-Lyrae, for example, are bright enough to be observed in the nearby galaxies, but, for more distant ones, we have to seek help from even brighter objects, like the supernovae.

In the last case, however, another problem arises. Supernovae are relatively rare, temporary and short-lived phenomena, so rare that just about 10 such events have occurred in the Milky Way during the last 2000 years, and none have happened recently, when their evolution could have been monitored and exploited for scientific uses. Actually, this is not necessarily a bad thing, as a supernova explosion might represent a serious threat to most living creatures (man included) if it happens too close to the Earth, but, in any case, it means that we have no parallax determination of the distance of supernovae, so this standard candle cannot be calibrated like the Cepheids.

But about two hundred galaxies lie within the "range of competence" of these variable stars, so astronomers were able to provide a calibration for the supernovae by means of the Cepheids, which, in their turn, were calibrated with the parallaxes. Even larger distances can be reached with other techniques, which rely on supernovae for calibration.

In this way, scientists have built an entire series of distance estimators, each with its own specific application range and each using the previous method for its calibration, quite appropriately christened the "cosmic distance ladder" (Fig. 10.10).

Thanks to the cosmic distance ladder, we know, for example, that quasars are far-off galaxies whose distance can be estimated in billions of light years, and that these huge "island-universes" are not distributed randomly, but rather gather in huge clusters, separated by even larger spaces. This gave origin to all of the questions about the formation and the hypothetical end of the universe that are still waiting for an answer.

The first step of our ladder, however, is the knowledge of the parallaxes that, once more, confirms their fundamental role in astronomy. Improving the quality and the range of stellar parallaxes automatically brings a series of chain improvements to the accuracy of the cosmic distance ladder, which, in its turn, is at the basis of practically every astrophysical investigation.

10.7 Fundamental Physics

Finally, the accuracy of Gaia's measurements could help to solve a problem mentioned in the third chapter, a problem that was born as a result of astrophysical investigations, but actually belongs to the realm of fundamental physics.

In that chapter, we talked about the two different problems of the missing mass and the dark energy, which involved a wide range of astrophysical phenomena, from the motion of stars within the galaxies to the cosmological expansion of the universe. Although the most supported explanation is one that involves the existence of a "dark side of the universe," in the form of dark matter and dark energy, many scientists are not satisfied with this so-called standard model.

The fact that we have to accept that dark components are needed to take 95% of the constituents of the universe into account, but that, at the same time, we have no evidence for them except those involving gravity, is quite disturbing. Scientists have been hunting for dark matter for more than 50 years, with no results, and dark energy was so unexpected that, after 20 years, it gets stuck into our equations in the form of a strange "field," but, in fact, physicists don't have a clear picture of what it could be.

It is not so strange, therefore, that somebody is asking whether all of these experimental evidences are a sign that gravity behaves differently from the way that we think.

The currently accepted theory that explains how gravity works is General Relativity, introduced by Albert Einstein more than a century ago. If we continue to fail in our efforts to understand the behavior of this interaction, then this is the theory that has to be overthrown, and one of the ways that we have to put General Relativity to the test is based on astrometric measurements.

Among other things, this theory states that light does not move in a straight line: its path is influenced by the masses of objects. A practical consequence is that, for example, if we look at a star when it happens to be visually close to the edge of the Sun, its position would change with respect to that which it has when its angular distance from the Sun is large. This phenomenon is dubbed the "gravitational bending" of the light.

It is beyond any doubt that an effect that influences the apparent position of stars in the sky can be fruitfully investigated with astrometry, and, in fact, the observation of the gravitational bending due to the Sun during a solar eclipse in 1919 provided striking experimental evidence in favor of General Relativity (Fig. 10.11).

Nevertheless, what once scored a point for Einstein might well, under different conditions, score a point against him. The astrometric measurements of the 1919 experiment were one hundred thousand times less precise than those of Gaia, and, at that time, General Relativity was proven right at a mere 10% level of accuracy. As of today, things have improved, and, with different techniques, scientists have improved the tests by a factor of ten thousand. Gaia, thanks to its incredible accuracy, will be able to improve this by another ten times.

The consequences can be striking. After General Relativity, several alternative theories of gravity have been formulated; many of them have already been

Fig. 10.11 One of the photographs of the solar eclipse of May 29, 1919, observed by the expedition of Dyson, Eddington and Davidson, which provided the first experimental evidence of the gravitational bending of light

disproved, but others still remain viable. Among the latter, many foresee deflections similar to, but quantitatively different from, those of Einstein's relativity; moreover, there are claims that such theories, which would change our understanding of the way gravity works, would explain the problem of the missing mass and that of the accelerating expansion of the universe without resorting to dark energy and dark matter!

Gaia, providing a precise measure of how light is deflected through its determination of stellar displacement, might tell us whether such an effect is better modelled by General Relativity or another theory. Whatever the result, its measurements will yield one of the best tests of gravity.

Recent research, moreover, suggests that Gaia can also help in the search for gravitational waves. Some years after the formulation of General Relativity, it was discovered that, according to this theory, variations in gravitational fields, like those produced, for example, by moving masses or collapsing objects, had to deliver energy that propagated through space-time, modifying its fabric.

This effect, however, is so tiny that it was measured experimentally only in 2015. A passing gravitational wave would cause the observed position of a star to oscillate, and it was thus suggested that Gaia's data could be sensitive to those produced by supermassive black holes, namely, the million- or billion-solar mass black holes that can frequently lie at the center of galaxies.

10.8 Recapping

We started this book from a very short summary of basic scientific knowledge, then we described, in some detail, the basic principles of astrometry. Successively, we entered into a more technical realm, explaining the functioning of the satellite, and how, from its raw measurements, we come to the processed data that are needed to fulfill all of the scientific expectations.

Today, Gaia advances along the path opened by the Hipparcos satellite, and its huge number of targets, combined with the high precision of this satellite, promise to foster great scientific results. We tried to explain them in this chapter by taking a virtual journey through the Milky Way and get a bird's eye view of the expectations of thousands of thrilled scientists; but will this mission be able to keep its promise?

With the release of the second intermediate version of the Gaia catalog, we already have a taste of the possibilities opened up by this mission, and, in the next part of this book, we will take a more detailed look at some of the current achievements.

Part V
Gaia's Latest News

And thence we came forth again
to see the stars.
Dante Alighieri, Inferno XXXIV, 139

11

Gaia Data Releases

11.1 Gaia Data Made Public

The European Space Agency decided that, to maximize the scientific return and public awareness of this project, not only must there be no proprietary rights to the mission data for the scientists who worked on it, but also that its results had to be made available to the entire scientific community—and to anybody who wishes to download them from the internet—even before the completion of the mission.

Obviously, these could not be the final results, but we have to remember that the Gaia catalog will be 1000 times larger than that of Hipparcos, and at least 1000 times more accurate than the other astrometric catalogs with a comparable number of objects. The value of the Gaia results, however, does not rest merely in the number of objects or the accuracy of the astrometric catalog, but rather in the new data that it will provide, together with the more "traditional" sorts, that is, stellar parallaxes never available before, supplemented by photometric and spectroscopic data. It is not the first time that, for example, proper motions have been estimated, nor is it the first time for parallaxes or physical characterization of the stars; but it is the first time that all of this, and much more, has been done to such a large extent, by a single measurement campaign, in an all-comprehensive and self-consistent way. Gaia delivers the complete package. Actually, the largest complete package ever delivered.

All of this makes clear why even some partial Gaia results, less complete and less mature than what is expected for the end of the mission, will surely be able to foster interesting scientific exploitation. It was also evident that a preliminary Gaia catalog supplying a more or less complete parallactic census across the Milky Way would have outmatched any of the previous catalogs.

© Springer Nature Switzerland AG 2019

G. Bernardi and A. Vecchiato, *Understanding Gaia*,
Springer Praxis Books, https://doi.org/10.1007/978-3-030-11449-7_11

As we have anticipated when speaking of the Gaia data reduction, the satellite scans the whole celestial sphere more or less every six months; this characteristic is exploited by so-called cyclic processing, which uses collections of months-long data stretches, called "data segments," to produce different versions of the catalog, or a "Gaia Data Release."

Although, in the DPAC's plans, a release is foreseen more or less every two or three years, they do not have a strictly regular cadence; rather, a catalog is made public when the DPAC deems it sufficiently appealing, both from the point of view of the scientific exploitation and that of its scientific reliability. When it happens, the data are published on the archive, along with a series of scientific papers. The size and complexity of Gaia data is such that a release could not be used without appropriate support for the end user. This is why, although data can even be downloaded as plain compressed csv files (basically, csv are just text files that can be opened by any text editor or a spreadsheet application), a number of tools are provided to help with scientific exploitation. They include, for example, an online search interface and engine that helps to filter and select data according to various criteria, summary statistics about the catalog, visualization tools able to produce on-the-fly plots for a preliminary exploration, and, last, but not least, extensive documentation that describes in full detail how the data have been processed and validated, their characteristics and the so-called "Data Model," an essential piece of information that describes the way that the data are structured, along with the meaning of each parameter included in the model.

At present, two data releases have been produced, the DR1 in 2016 and the DR2 in 2018. The next one will probably be delivered in 2021, but the total number is still to be determined, and it might change according to the actual duration of the mission. Indeed, Gaia was foreseen to end its operational phase after 5 years of measurements, in 2019, but the ESA has already granted a first extension through 2020, and more might be decided upon in the future, limited by the duration of the propellant needed for the attitude control and the orbital maneuvers.

11.2   Data Release 1

The first Gaia Data Release, or DR1, was officially delivered on September 14, 2016, less than three years from the launch and about a couple of years after the end of the commissioning phase. Although its scientific value, with respect to the final goal of the mission, was still limited, it represented an important milestone in many respects.

In fact, it served as a testbed for both the DPAC and the external scientific community. The former could exercise its capability of producing a data release, thus testing, for the first time, the whole processing chain, from the satellite to the archive, and was able to provide a useful early calibration and internal validation of the instrument, whereas the latter could get familiar with the Gaia data and the archive tools as early as a few years after the launch, even before half of the mission's duration had passed. Moreover, regarding the scientific value of the data, one has to remind oneself of what was mentioned in the previous section: even an incomplete release will provide astronomers with a lot of work and a lot of fun.

The level of interest raised by an experiment, or its scientific productivity, can be estimated by the number of related scientific publications, so the above statement can be quickly proven true by taking a look at publicly available databases of scientific papers. Considering only those using the DR1 data, that is, neglecting those that only refer to Gaia in general, or citing discoveries made by Gaia, one can easily find at least a hundred of papers in ADS, the most popular bibliographic database used by the astronomical community, published in a period of just about one year after the release.

This release used the observations from the first two segments: segment 00, spanning from July 25, 2014, to June 4, 2015, and segment 01, from the end of the previous one to September 16, 2015. Already at this stage, an astrometrist can tell that choosing to deliver a release with these sets of data looks strange, at least for a mission whose primary target is to provide a 3D map of the sky.

The main problem, in fact, is that less than 14 months of measurements is too short a period to estimate the promised parallaxes and proper motions, and, indeed, this data release was not created using Gaia data alone; actually, it required a little "trick," and the help of the results of its predecessor, Hipparcos (Fig. 11.1).

Proper motions and parallaxes play together in changing the position of a star, the latter with a periodic wobble and the former with a "straight" motion, and since the parallax is due to the orbital motion of the Earth around the Sun, the period of the parallactic oscillation is one year. This has to be disentangled from the proper motion, and, for this reason, we have to follow the object for a period of time significantly longer than a single year, in order to separate the two effects (Fig. 11.2).

For the DR1, the DPAC scientists used the positions of Tycho-2, one of the two Hipparcos catalogues; this catalog could also supply its own estimation of the proper motion, and the other Hipparcos catalog might also have provided parallaxes as additional help. Except that these data would have not helped at all, and there was no need to make use of anything other than the positions. The trick worked because, just by comparing the positions of Gaia with those of Tycho-2, which were more than 20 years older, it was possible to estimate the proper motions independently. Finally, knowing the proper motion, its contribution could

Fig. 11.1 Artistic impression of the Hipparcos satellite (© ESA)

be separated from that of the parallax in the Gaia measurements, which eventually became measurable. This is obviously a very simplified explanation of the actual procedure, but it clarifies the basic concept used for solving this apparently unsolvable problem.

The one limitation was that Tycho-2 contains only a little more than 2 million stars, and therefore the DR1 catalog gives all 5 astrometric parameters for these stars alone, while, for the other 1.1 billion, it provides only their positions.

The DR1 is formed by two more datasets, both delivering photometric information, that is, magnitudes. The first one, however, is a catalog of mean magnitudes for all of the sources, while the second regards stars that show a variable luminosity. It offers the light curves, that is, the temporal variation of the magnitude of about 3000 Cepheids and RR-Lyrae, special types of star that we mentioned in the first part for their importance as standard candles that help us to climb up to the next step of the cosmic distance ladder.

Obviously, some limitations resulting from the short leg of the mission were unavoidable. For example, the catalog was, in many ways, incomplete, because of such factors as missing faint and bright stars, missing high-proper motion stars,

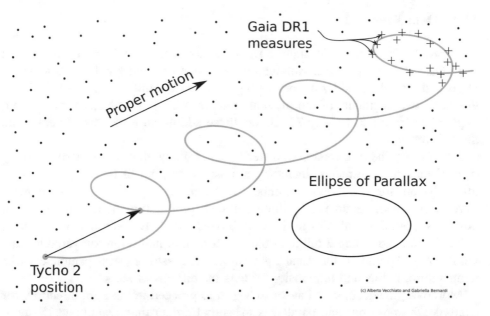

Fig. 11.2 The grey curve shows the path of a star over a background of distant stars as it results from the combination of its proper motion and parallax. DR1 measurements, represented by crosses in the upper right-hand corner, were combined with the Tycho-2 positions to estimate both proper motion and parallax. The picture is not a rigorous representation of the actual situation, because each loop takes one year, and Tycho-2 positions are about 20 years older than those in DR1 (*Credits* Alberto Vecchiato and Gabriella Bernardi)

and subpar coverage of some regions of the sky along the ecliptic. Another problem is in the presence of so-called systematic errors, that is, offsets in the estimated parameters whose origin is different from a pure random uncertainty. These are particularly dangerous because they can bias other results. The most dangerous of these systematics is in regard to the parallaxes, which are estimated to have a bias of about 0.1 mas. Finally, this catalog shows other issues in the photometric data, involving clearly incorrect magnitude values, or faulty estimation of the magnitude uncertainties.

All of these problems were clearly reported in the documentation, because they had to be taken into account for the correct use of Gaia's DR1 data and in the interpretation of the consequent results. However, the situation has significantly improved with the introduction, two years later, of the second Gaia Data Release.

11.3   Data Release 2

On April 25, 2018, the ESA and the DPAC announced the delivery of the second Gaia Data Release, or DR2. Although it came just one and a half years after the DR1, and was obtained by the processing of only 22 months of data—it includes the same two segments of the previous release, plus segment 2, spanning from September 16, 2015, to May 23, 2016—these additional 8 months of data made the difference.

It is, in fact, the first version of a "real" Gaia catalog, that is, one derived solely from Gaia's measurements, including almost every kind of data expected from this mission for the "right" number of sources. So, we have 5-parameter astrometry for more than 1.3 billion sources (plus a further 361 million with positions alone), magnitudes in the visible band for all 1.7 billion sources, radial velocities for more than 7 million objects, more than half a million variable stars with almost 4 hundred thousand light curves, and astrophysical parameters like temperature, radius and luminosity for tens of millions of stars.

Moreover, the accuracies have also begun to proceed in the right direction, for example, the positions and parallaxes for stars brighter than magnitude 15 are at the 20–40 µas level, while those of magnitude 20 have a typical uncertainty of 700 µas.

Despite these great advances, as with the previous catalog, scientists who wish to use these data have to be aware of its limitations. It is, in fact, inevitable that, after only 22 months, the data processing still needs to be improved in some respects. In part, this is obviously due to the lack of data, and, in part, it also comes from the need to obtain better calibrations that can cure the problems arising from an imperfect knowledge of the instrument.

There is still a problem of completeness at both ends of the magnitude scale, but this is much less significant with respect to the DR1. As strange as it might seem, for the Gaia instrument, stars of magnitude 7 or brighter are quite difficult to measure, because its "eyes" are designed to be very sensitive and to catch small quantities of light. In this way, the satellite's instrumentation can detect sources fainter than magnitude 20, but, at the same time, objects invisible to our eyes are like powerful headlights that can easily blind the delicate Gaia detectors. More-over, objects fainter than magnitude 17 may be missing, because there are so many of them. When Gaia scans crowded regions like those of the galactic disc, they can be so numerous that, simply put, they have to be filtered out to allow the satellite to keep up its processing.

Systematic errors are now smaller, but they still exist, in particular, for the parallaxes and for photometric parameters, and radial velocities, especially the large ones, have to be used with some precaution.

Another important addition to the Gaia catalog is a dataset of astrometric and photometric data for a sample of solar-system bodies. It comprises 14,099 objects, a number that seems to vanish, compared to the intimidating millions or billions of previous lines; nonetheless, it is the first time that objects of this kind are included in the Gaia catalog, and the situation will improve with future releases.

Last, but not least, contrary to objects in the solar system, one important class of objects that is still missing is that of the exoplanets, those whose reduction pipeline, in fact, requires a more refined calibration to give meaningful results; we will have to wait for the future releases to see them in the Gaia catalog.

12

Selected Results

12.1 Hunting for Science

The analysis of Gaia's data, and the exploitation of its catalogs by the scientific community, has already generated many interesting results. The scientific publications that have originated from this mission number in the hundreds, even if counting only those that came out after the results of just 22 months of the total 5 years (now at least 6) of the mission had been published.

These studies and discoveries involve diverse astronomical fields, from stellar to galactic evolution, but we have been pleasantly surprised by some of the unexpected results regarding exoplanetary and cosmological research as well. It was estimated, in fact, that no news would be heard from this front, because the detection of exoplanets requires a high-precision Gaia catalog and the gathering of a long temporal stretch of data.

We can thank the fantasies of the astronomers, as well as their impatience and eagerness for producing new discoveries, if we end up having such a rich scientific harvest so early. In many cases, in fact, interesting results were obtained by combining the data from Gaia with those from other catalogs, or coming from other instruments, in a fruitful collaboration with other projects.

Those that we are presenting in this chapter represent just a very limited selection of a few of the most interesting discoveries, and, obviously, the choice can be biased by the personal taste of the authors, as well as by an incomplete survey of the more than one thousand papers related to the various aspects of this mission that can be found in the scientific literature of the last three years alone.

© Springer Nature Switzerland AG 2019

G. Bernardi and A. Vecchiato, *Understanding Gaia*,
Springer Praxis Books, https://doi.org/10.1007/978-3-030-11449-7_12

12.2   Stellar NICUs Are Less Exclusive Than Expected

Stars are born from the collapse of huge nebulae of gas and dust, driven by the gravitational force, and it was generally believed that these births always take place in large groups, forming dense associations of stars called "stellar clusters."

These should not be confused with globular clusters, which are those extremely dense, spherical objects orbiting around the center of the Milky Way, mostly formed of old stars, which we wrote about in the second chapter when we briefly accounted for the early investigations into the nature of our galaxy. Rather, we are dealing with different clusters, located inside of the Milky Way and called "open clusters," which come in more irregular shapes and are formed by young stars.

Since we observe stars scattered across the whole Milky Way at very different levels of clustering, from those that are very dense to those that are very loose, this scenario has to explain why, once they start their life together, the newly formed stars eventually reach legal age and depart from their nest, going their own way and living, for the most part, a lonely life, or in tight ensembles of a few stars, all orbiting around each other.

The most popular model was based on the hypothesis that clusters could lose a large part of their gas shortly after the star formation process begins. These giant associations are kept bounded by gravity, which depends on the amount of mass contained in the cluster, so, if the cluster loses a significant share of its free-gas component, it might well be that the remaining mass is not sufficient to keep the newborn stars together, and they can start their wandering lives, disrupting the original association.

This scenario, therefore, needed, in the first place, a robust mechanism that could explain the loss of gas, and this was cleverly found in an event that fits the model very well, as it is characteristic of the first stages of stellar formation. A star is born when it starts burning hydrogen with the thermonuclear fusion in its core, but this ignition generates lots of energy and lots of photons that start to irradiate from the stellar surface; these photons generate a violent "wind" that pushes away the surrounding gas and dust, clearing the planetary nebula.

Thus, such a mechanism, summing up the contributions of several newborn stars, was deemed sufficient to take into account the required gas loss. The resulting model —which came in two varieties, known as "singularly monolithic" and "multiply monolithic," associated with just one original cluster or with several nearby clusters —was supposed to explain the whole story, and the clusters, like gigantic "stellar NICUs," were regarded as the only place in which stars could form, but the picture has changed a lot in recent years, also thanks to results from Gaia's data.

For example, astronomers started to argue that the efficiency of star formation is much higher than what was previously thought, so that newborn stars could ingest almost all of the free gas before starting to generate the stellar wind, and thus the amount of gas lost with this mechanism could not have a significant effect on the overall gravity of the cluster. Moreover, an alternative model was proposed, arguing that stellar formation could happen at different ranges of stellar density, and thus not exclusively in the clusters.

Two scientists from Heidelberg University, Germany, recently approached this problem with the help of the Gaia data from DR1. In particular, they used the parallaxes and proper motions that are available for the subset of the solution obtained for the Tycho-2 stars to identify the members of 18 so-called "OB associations." These are very high-massive stars of spectral types O and B, which do not live long enough to have time to move very far from their place of origin. For this reason, we can often find them in groups, the associations.

If the monolithic model is correct, then the associations should show a sign of an expansion from their original cluster, whereas lack of detection of such a sign would be evidence in favor of the other model. The astronomers, then, used the same Gaia data to compute a series of parameters measuring the degree to which these associations are coming from a previously denser state, but none of them showed any evidence in this sense. Rather, the estimators suggested that the stars were moving randomly, exactly what should be expected if they were formed according to the alternative model.

12.3   Secrets of Stellar Structures Revealed by a Tiny Gap

We have repeatedly mentioned that, in Astronomy, the Hertzsprung-Russel diagram is a fundamental tool with many applications. One of the recent mission discoveries is a clear illustration of the truth of this assertion. This is also an example of a discovery that was made possible only thanks to the sheer power of the size of the Gaia catalog.

With the impressive number of about 1.5 billion stars in its second release, the DPAC scientists could build the densest and most accurate version of the H-R diagram ever realized, which, upon careful examination by a team of astronomers from the USA and UK, revealed a tiny "gap" in the region populated by the so-called red-dwarfs (Fig. 12.1), small stars with a mass about one-third that of the Sun.

The gap is so subtle that it looks like a small and almost imperceptible scratch on the otherwise continuous distribution of stars on the graph, and it can surely only be noticed because of the high density number of stars in this version of the diagram, but it is definitely real. Once discovered in the DR2 catalog, in fact, researchers looked for it in other ones, which had fewer stars but were looking in different bands.

The accurate astrometry of the Gaia catalog made it possible to "cross-match" the different catalogs, that is, to identify the stars in the Gaia catalog that are

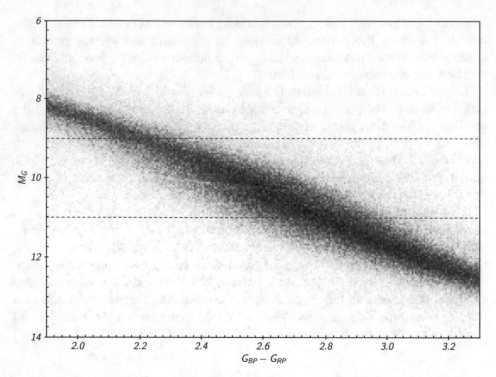

Fig. 12.1 Close-up of the HR diagram created with Gaia's data showing the narrow gap around magnitude $M_G = 10$ (Courtesy Wei-Chun Jao, Georgia State University)

members of the other one, so it was possible to build other H-R diagrams using photometric information from different bands, and they found that this feature was still present. This allows us to cast away the hypothesis that the gap was due to a problem in the Gaia data. Later, another work, by astronomers from the University of Delaware, found a plausible explanation for such a gap in terms of the physics of stellar interiors.

When a star starts burning its thermonuclear fuel, it generates a lot of energy, which has to be transferred from its core to the surface. The way this happens depends on the mass of the star.

For the smallest of the red dwarfs of spectral type M, this process looks like boiling water in a pot; the hotter plasma in the center moves up to the stellar surface, where it cools down, and thus falls towards the center, where it heats up again, starting another cycle (Fig. 12.2, left). Technically speaking, such a mechanism is called convection, and stars like these are said to have a totally convective interior.

However, when the stellar mass exceeds a certain limit—that is, when it is larger than about 0.5 solar masses—it is no longer possible to dispose of the internal

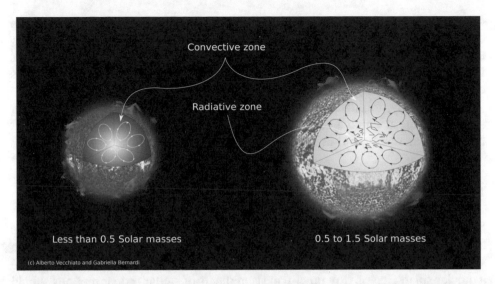

Fig. 12.2 In their mature stage, stars with a mass smaller than 0.5 solar masses behave like boiling pots of water, disposing of the excess heat by convective motion, whereas those with a larger mass have a radiative core surrounded by a convective envelope (*Credits* Alberto Vecchiato and Gabriella Bernardi)

energy with a convective mechanism alone. The pressure of the total mass is so large at the center that it prevents the needed motion in the core, and convective transfer can happen only from a certain distance from the center. Energy, thus, escapes from the inner region thanks to the photons, which can take hundreds of thousands of years to reach the outer convective region (Fig. 12.2, right).

Moreover, stellar luminosity and color changes with age, because of the change in the energy sources due to the burning of their fuel by nuclear reactions. This implies that a star moves in the H-R diagram as time goes by, and the probability of finding objects in certain positions of the diagram depends on the way in which stars evolve their colors and luminosities over time.

For smaller stars, the nuclear fuel changes with time, but the convective motion is very efficient in transporting the different kinds of matter across the stellar interior, and the nuclear reactions can change in a smooth way, so that their luminosity steadily increases with time.

Gaia, instead, allowed us to search for a narrow interval of masses, between 0.315 and 0.345 solar masses, in which the star experiences an evolution in its interior. In fact, for a certain amount of time, which depends on their mass, they have a convective core and a convective envelope separated by a radiative zone. Therefore, those stars rely both on the convective and on the radiative mechanisms, but the latter is less efficient in intermixing the nuclear fuel, and we have a status of equilibrium during this period that keeps the luminosity more or less constant. When the composition of the stellar interior changes and the equilibrium cannot stand anymore, the inner and outer convective regions merge, and the star, now fully convective, can burn its fuel with more efficiency, increasing its luminosity at a steady pace. It is this period of equilibrium that makes these stars cover a narrower range of luminosities, creating the gap.

Contrastingly, for stars with a mass larger than 0.35 solar masses, they never reach the stage at which the merging happens. These stars, therefore, increase their luminosity smoothly, but at a slower rate than their smaller-mass fellows.

12.4   Shedding Light on the Milky Way's Turbulent Rendezvouses

Scientists have always expected that Gaia would be able to tell us many things about the evolution of the Milky Way, and, actually, this was one of the main scientific drivers in favor of the mission. In the previous part, we also tried to give a quick explanation of why precise measurements of stellar proper motions should give a detailed accounting of the dynamical history of our galaxy, and here we report on how, already in its early stages, Gaia duly met our expectations. Until now, in fact, astronomers have been able to prove at least two collisions between the Milky Way and another galaxy.

The first one was a major impact, or—since we are not speaking of compact objects, but of large stellar ensembles—a major merge between our galaxy and another one, dubbed "The sausage galaxy" or "The Gaia sausage." This object is named after the plot that made its discovery possible. Indeed, when the team of discoverers tried to plot the velocities of 7 million stars from the DR2 catalog, separating their radial motion from the circular one, two very evident groups formed in their graph; one had exactly the shape that should be expected for stars orbiting around the galactic center in circular orbits, whereas another one showed evident signs of very elongated orbits, with a large radial component. In the plot, it looks like a sausage, with its long side along the radial axis (Fig. 12.3).

Calculations showed that the collision happened between 8 and 11 billion years ago, and that it involved a dwarf galaxy of more than 10 billion solar masses. This is much smaller than the Milky Way, but, having approximately the size of the Large Magellanic Cloud, it is also larger than typical mergers. Actually, it was so large that it had its own globular clusters, eight of which are now clusters of the Milky Way.

The second collision, however was much less dramatic, but it also happened much more recently. Astronomers already knew that our nearest galactic neighbor, the Sagittarius dwarf galaxy, has a long history of repeated collisions with the Milky Way, which, according to some theories, contributed to the creation of the spiral structure of our galaxy. This object is believed to have a polar orbit around

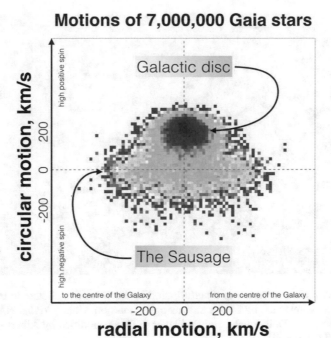

Fig. 12.3 Plotting the motions of 7 million Gaia stars put to evidence of the presence of a sausage-like shape generated by stars with high radial motion distribution coming from an ancient galactic collision (*Credits* V. Belokurov (Cambridge, UK) and Gaia/ESA)

our galaxy with a period of several hundred million years, and its last pass by the galactic disk, between 300 and 900 million years ago, caused a great deal of perturbations in its stars, even if it was not a direct hit. Its signs, apparently, are still visible today in the data of the Gaia DR2 catalog.

Astronomers, in fact, have examined the motion of a sample of about 6 million disk stars, and they observed a large number of structures, described as similar to the "ripples and waves" that form in the water when we throw a stone into a pond (Fig. 12.4). The analysis of these structures showed that they formed in approximately the same period as the last graze of Sagittarius with the Milky Way.

12.5 Tinkering with the Cosmic Distance Ladder

When, in a previous chapter, we mentioned the various techniques that can be used to estimate cosmic distances, we failed to mention the last step of the cosmic distance ladder, namely, the method based on the measurement of the redshift of the observed objects, which allows us to reach the most remote parts of the

Fig. 12.4 Artist's impression of a perturbation in the velocities of stars in our Galaxy, the Milky Way, revealed by ESA's star mapping mission, Gaia (*Credits* NASA, ESA, and F. Summers, J. DePasquale, and D. Player (STScI), modified by Alberto Vecchiato and Gabriella Bernardi)

universe. We already know what the redshift (or the blueshift) is, because it has been shown why it is used by Gaia—and in astronomy in general—to measure the radial velocity of astronomical objects.

Moreover, in one of the first chapters, the one dealing with galaxies, we learned that Hubble discovered a peculiar phenomenon that connects the receding speed of distant galaxies with their distance. More precisely, we said that these two quantities—the speed of recession and the distance—are proportional to each other, so that the farther the object, the faster it moves away from us.

We can estimate the receding speed of an object from its redshift, thus, if one knows the constant of proportionality, it would be easy to deduce its distance using the Hubble law that relates speeds with distances. As you can imagine, however, knowing the constant of proportionality—named the Hubble parameter and indicated in the formulae with the letter H—is anything but simple. Indeed, Hubble's first estimation, in 1929, gave a value of about 500 km/s/Mpc, which means that, after 1 million parsecs, the receding speed is 500 km/s faster, which is about 7 times greater than the currently accepted value or, for that matter, as we are going to discover in a while, than the two currently competing values!

First of all, calling H "constant" is rather misleading, and we have to understand exactly what it is that astronomers are estimating. From cosmological theories, in fact, the velocity we deduce from it has not been intended in the normal sense to mean how much space is covered by a source in a given amount of time, but rather as a measure of how the space itself is expanding. The redshift effect looks the same, but only because the wavelength of the light is "stretched" by the expanding space. So, the value of H depends on the speed of this expansion process, which, however, is governed by gravity, and therefore this parameter is not constant in time. On the other hand, the Hubble parameter has a unique value for each given time, so that it is "constant in space," and astronomers estimate its value at current time, which is indicated by the symbol H_0.

Second, as we have just mentioned, this relation is valid only for very distant objects and cannot be used, for example, for the stars in our galaxy. In other words, this is an indicator for cosmological distances, and thus H_0 can only be determined by knowing the distance of some very distant objects in the first place. This is the very reason why the redshift method is the last rung of the cosmic distance ladder.

As happened with the other methods, then, first one needs to find some standard candles in cosmologically distant objects, like a supernova. Once we know their distance, we can use Hubble's law and their measured redshifts to estimate the value of H_0.

But the supernovae are just the penultimate rung, and have to be calibrated in their turn, until we get back to the first method of estimating cosmic distances, that is, the parallax. So, the accuracy of the parallaxes has a knock-on effect on those of the other subsequent techniques, up to the determination of the Hubble constant, and that is exactly what happened recently.

A group of US and Italian scientists set out to use the breakthrough advance represented by the Gaia parallaxes to adjust all of the rungs of the cosmic distance ladder. Basically, the team exploited the new and improved parallaxes of 50 Milky Way Cepheids made available by the DR2 data to obtain a better calibration of the Period-Luminosity-Color relation that makes these objects useful as standard candles, and climbed the ladder up to the Hubble constant, as described above. The other ingredient they needed for this task was a precise determination of the brightness of these objects with respect to their extragalactic counterparts, which was provided by photometric measurements made by the Hubble Space Telescope. In this way, the value of H_0 was estimated with just a 2% accuracy, and it is expected that, with the final results of Gaia, this result could be improved to the 1% accuracy level.

But this is not the end of the story, and instead of giving a final answer, this fine measurement contributed to the confirmation of a puzzling mystery about the actual value of the Hubble constant. Indeed, the value obtained by the previous team is 73.52 km/s/Mpc, with an uncertainty of 1.62 km/s/Mpc, but other values have been obtained with other independent techniques and other measurements.

Among those others, the most precise one is that coming from the ESA Planck mission, which found a value of 67.66 ± 0.42 km/s/Mpc, an estimation that is clearly incompatible with the previous one. The Planck scientists obtained this value by creating a completely different map from that of Gaia. In fact, they mapped the universe as it was about 350,000 years after the Big Bang, looking at the so-called Cosmic Microwave Background—which represents the remnant of the moment in which the photons started to propagate freely across the universe—and deducing from its characteristics an entire set of physical quantities, including H_0.

Explaining this difference is a great challenge for science, but it could result in a breakthrough of our understanding of the universe and its evolution.

References

Books

Bernardi, Gabriella. *La Galassia di Gaia*, Edizioni la Ricotta, 2013 (In Italian)
Bernardi, Gabriella. *The Unforgotten Sisters - Female Astronomers and Scientists before Caroline Herschel*, Springer, 2016
Green, Robin M. *Spherical astronomy*, Cambridge University Press, 1985
Kartunnen, Hannu. et al. Fundamental Astronomy, Springer, 2007
Keating, Brian. *Losing the Nobel Prize*, W. W. Norton and Company, Inc., 2018
Wynn-Williams, Gareth. *Surveying the Skies*, Springer, 2016

Scientific Papers

Antoja, T., 2018. A dynamically young and perturbed Milky Way disk. *Nature*, **561**, 360
Belokurov, V. et al., 2018. Co-formation of the disc and the stellar halo. *Monthly Notices of the Royal Astronomical Society*, **478**, 611
Gaia Collaboration, 2016. The Gaia mission. *Astronomy & Astronpysics*, **595**, A1
Gaia Collaboration, 2016. Gaia Data Release 1: Summary of the astrometric, photometric, and survey properties. *Astronomy & Astronpysics*, **595**, A2
Gaia Collaboration, 2018. Gaia Data Release 2: Summary of the contents and survey properties. *Astronomy & Astronpysics*, **616**, A1
MacDonald, J. and Gizis, J., 2018. An explanation for the gap in the Gaia HRD for M dwarfs. arXiv:1806.11454v3 preprint (https://arxiv.org/abs/1806.11454), to appear in *Monthly Notices of the Royal Astronomical Society*
Moitinho, A. et al., 2017. Gaia Data Release 1: The archive visualisation service. *Astronomy & Astronpysics*, **605**, A52

© Springer Nature Switzerland AG 2019
G. Bernardi and A. Vecchiato, *Understanding Gaia*,
Springer Praxis Books, https://doi.org/10.1007/978-3-030-11449-7

Riess, Adam G. et al., 2018. Milky Way Cepheid Standards for Measuring Cosmic Distances and Application to Gaia DR2: Implications for the Hubble Constant. *The Astrophysical Journal*, **861**, 126

Ward, Jacob L. and Kruijssen, Diederik J. M., 2018. Not all stars form in clusters – measuring the kinematics of OB associations with Gaia. *Monthly Notices of the Royal Astronomical Society*, **475**, 5659

Wei-Chun Jao et al., 2018. A Gap in the Lower Main Sequence Revealed by Gaia Data Release 2. *The Astrophysical Journal Letters*, **861**, L11

Web

ESA Science Portal, http://gaia.esa.int

ESA Science and Technology portal, http://sci.esa.int/gaia/

ESA website for the scientific community, https://www.cosmos.esa.int/web/gaia

Gaia Archive, https://gea.esac.esa.int/archive/

Gaia pages of the Earth Observation Portal, https://directory.eoportal.org/web/eoportal/satellite-missions/g/gaia